Amphibians: A Very Short Introduction

VERY SHORT INTRODUCTIONS are for anyone wanting a stimulating and accessible way into a new subject. They are written by experts, and have been translated into more than 45 different languages.

The series began in 1995, and now covers a wide variety of topics in every discipline. The VSI library currently contains over 650 volumes—a Very Short Introduction to everything from Psychology and Philosophy of Science to American History and Relativity—and continues to grow in every subject area.

Very Short Introductions available now:

ABOLITIONISM Richard S. Newman
THE ABRAHAMIC RELIGIONS Charles L. Cohen
ACCOUNTING Christopher Nobes
ADAM SMITH Christopher J. Berry
ADOLESCENCE Peter K. Smith
ADVERTISING Winston Fletcher
AERIAL WARFARE Frank Ledwidge
AESTHETICS Bence Nanay
AFRICAN AMERICAN RELIGION Eddie S. Glaude Jr
AFRICAN HISTORY John Parker and Richard Rathbone
AFRICAN POLITICS Ian Taylor
AFRICAN RELIGIONS Jacob K. Olupona
AGEING Nancy A. Pachana
AGNOSTICISM Robin Le Poidevin
AGRICULTURE Paul Brassley and Richard Soffe
ALBERT CAMUS Oliver Gloag
ALEXANDER THE GREAT Hugh Bowden
ALGEBRA Peter M. Higgins
AMERICAN BUSINESS HISTORY Walter A. Friedman
AMERICAN CULTURAL HISTORY Eric Avila
AMERICAN FOREIGN RELATIONS Andrew Preston
AMERICAN HISTORY Paul S. Boyer
AMERICAN IMMIGRATION David A. Gerber
AMERICAN LEGAL HISTORY G. Edward White
AMERICAN MILITARY HISTORY Joseph T. Glatthaar
AMERICAN NAVAL HISTORY Craig L. Symonds
AMERICAN POLITICAL HISTORY Donald Critchlow
AMERICAN POLITICAL PARTIES AND ELECTIONS L. Sandy Maisel
AMERICAN POLITICS Richard M. Valelly
THE AMERICAN PRESIDENCY Charles O. Jones
THE AMERICAN REVOLUTION Robert J. Allison
AMERICAN SLAVERY Heather Andrea Williams
THE AMERICAN SOUTH Charles Reagan Wilson
THE AMERICAN WEST Stephen Aron
AMERICAN WOMEN'S HISTORY Susan Ware
AMPHIBIANS T. S. Kemp
ANAESTHESIA Aidan O'Donnell
ANALYTIC PHILOSOPHY Michael Beaney
ANARCHISM Colin Ward
ANCIENT ASSYRIA Karen Radner
ANCIENT EGYPT Ian Shaw
ANCIENT EGYPTIAN ART AND ARCHITECTURE Christina Riggs
ANCIENT GREECE Paul Cartledge
THE ANCIENT NEAR EAST Amanda H. Podany
ANCIENT PHILOSOPHY Julia Annas

ANCIENT WARFARE Harry Sidebottom
ANGELS David Albert Jones
ANGLICANISM Mark Chapman
THE ANGLO-SAXON AGE John Blair
ANIMAL BEHAVIOUR Tristram D. Wyatt
THE ANIMAL KINGDOM Peter Holland
ANIMAL RIGHTS David DeGrazia
THE ANTARCTIC Klaus Dodds
ANTHROPOCENE Erle C. Ellis
ANTISEMITISM Steven Beller
ANXIETY Daniel Freeman and Jason Freeman
THE APOCRYPHAL GOSPELS Paul Foster
APPLIED MATHEMATICS Alain Goriely
ARBITRATION Thomas Schultz and Thomas Grant
ARCHAEOLOGY Paul Bahn
ARCHITECTURE Andrew Ballantyne
ARISTOCRACY William Doyle
ARISTOTLE Jonathan Barnes
ART HISTORY Dana Arnold
ART THEORY Cynthia Freeland
ARTIFICIAL INTELLIGENCE Margaret A. Boden
ASIAN AMERICAN HISTORY Madeline Y. Hsu
ASTROBIOLOGY David C. Catling
ASTROPHYSICS James Binney
ATHEISM Julian Baggini
THE ATMOSPHERE Paul I. Palmer
AUGUSTINE Henry Chadwick
AUSTRALIA Kenneth Morgan
AUTISM Uta Frith
AUTOBIOGRAPHY Laura Marcus
THE AVANT GARDE David Cottington
THE AZTECS Davíd Carrasco
BABYLONIA Trevor Bryce
BACTERIA Sebastian G. B. Amyes
BANKING John Goddard and John O. S. Wilson
BARTHES Jonathan Culler
THE BEATS David Sterritt
BEAUTY Roger Scruton
BEHAVIOURAL ECONOMICS Michelle Baddeley
BESTSELLERS John Sutherland
THE BIBLE John Riches
BIBLICAL ARCHAEOLOGY Eric H. Cline
BIG DATA Dawn E. Holmes
BIOCHEMISTRY Mark Lorch
BIOGEOGRAPHY Mark V. Lomolino
BIOGRAPHY Hermione Lee
BIOMETRICS Michael Fairhurst
BLACK HOLES Katherine Blundell
BLOOD Chris Cooper
THE BLUES Elijah Wald
THE BODY Chris Shilling
THE BOOK OF COMMON PRAYER Brian Cummings
THE BOOK OF MORMON Terryl Givens
BORDERS Alexander C. Diener and Joshua Hagen
THE BRAIN Michael O'Shea
BRANDING Robert Jones
THE BRICS Andrew F. Cooper
THE BRITISH CONSTITUTION Martin Loughlin
THE BRITISH EMPIRE Ashley Jackson
BRITISH POLITICS Tony Wright
BUDDHA Michael Carrithers
BUDDHISM Damien Keown
BUDDHIST ETHICS Damien Keown
BYZANTIUM Peter Sarris
C. S. LEWIS James Como
CALVINISM Jon Balserak
CANADA Donald Wright
CANCER Nicholas James
CAPITALISM James Fulcher
CATHOLICISM Gerald O'Collins
CAUSATION Stephen Mumford and Rani Lill Anjum
THE CELL Terence Allen and Graham Cowling
THE CELTS Barry Cunliffe
CHAOS Leonard Smith
CHARLES DICKENS Jenny Hartley
CHEMISTRY Peter Atkins
CHILD PSYCHOLOGY Usha Goswami
CHILDREN'S LITERATURE Kimberley Reynolds
CHINESE LITERATURE Sabina Knight
CHOICE THEORY Michael Allingham
CHRISTIAN ART Beth Williamson
CHRISTIAN ETHICS D. Stephen Long

CHRISTIANITY Linda Woodhead
CIRCADIAN RHYTHMS
Russell Foster and Leon Kreitzman
CITIZENSHIP Richard Bellamy
CITY PLANNING Carl Abbott
CIVIL ENGINEERING
David Muir Wood
CLASSICAL LITERATURE William Allan
CLASSICAL MYTHOLOGY
Helen Morales
CLASSICS Mary Beard and
John Henderson
CLAUSEWITZ Michael Howard
CLIMATE Mark Maslin
CLIMATE CHANGE Mark Maslin
CLINICAL PSYCHOLOGY
Susan Llewelyn and
Katie Aafjes-van Doorn
COGNITIVE NEUROSCIENCE
Richard Passingham
THE COLD WAR Robert J. McMahon
COLONIAL AMERICA Alan Taylor
COLONIAL LATIN AMERICAN
LITERATURE Rolena Adorno
COMBINATORICS Robin Wilson
COMEDY Matthew Bevis
COMMUNISM Leslie Holmes
COMPARATIVE LITERATURE
Ben Hutchinson
COMPLEXITY John H. Holland
THE COMPUTER Darrel Ince
COMPUTER SCIENCE Subrata Dasgupta
CONCENTRATION CAMPS
Dan Stone
CONFUCIANISM Daniel K. Gardner
THE CONQUISTADORS
Matthew Restall and
Felipe Fernández-Armesto
CONSCIENCE Paul Strohm
CONSCIOUSNESS Susan Blackmore
CONTEMPORARY ART
Julian Stallabrass
CONTEMPORARY FICTION
Robert Eaglestone
CONTINENTAL PHILOSOPHY
Simon Critchley
COPERNICUS Owen Gingerich
CORAL REEFS Charles Sheppard
CORPORATE SOCIAL
RESPONSIBILITY Jeremy Moon
CORRUPTION Leslie Holmes
COSMOLOGY Peter Coles
COUNTRY MUSIC Richard Carlin
CREATIVITY Vlad Glăveanu
CRIME FICTION Richard Bradford
CRIMINAL JUSTICE Julian V. Roberts
CRIMINOLOGY Tim Newburn
CRITICAL THEORY
Stephen Eric Bronner
THE CRUSADES Christopher Tyerman
CRYPTOGRAPHY Fred Piper and
Sean Murphy
CRYSTALLOGRAPHY A. M. Glazer
THE CULTURAL REVOLUTION
Richard Curt Kraus
DADA AND SURREALISM
David Hopkins
DANTE Peter Hainsworth and
David Robey
DARWIN Jonathan Howard
THE DEAD SEA SCROLLS
Timothy H. Lim
DECADENCE David Weir
DECOLONIZATION Dane Kennedy
DEMENTIA Kathleen Taylor
DEMOCRACY Bernard Crick
DEMOGRAPHY Sarah Harper
DEPRESSION Jan Scott and
Mary Jane Tacchi
DERRIDA Simon Glendinning
DESCARTES Tom Sorell
DESERTS Nick Middleton
DESIGN John Heskett
DEVELOPMENT Ian Goldin
DEVELOPMENTAL BIOLOGY
Lewis Wolpert
THE DEVIL Darren Oldridge
DIASPORA Kevin Kenny
DICTIONARIES Lynda Mugglestone
DINOSAURS David Norman
DIPLOMACY Joseph M. Siracusa
DOCUMENTARY FILM
Patricia Aufderheide
DREAMING J. Allan Hobson
DRUGS Les Iversen
DRUIDS Barry Cunliffe
DYNASTY Jeroen Duindam
DYSLEXIA Margaret J. Snowling
EARLY MUSIC Thomas Forrest Kelly
THE EARTH Martin Redfern

EARTH SYSTEM SCIENCE Tim Lenton
ECOLOGY Jaboury Ghazoul
ECONOMICS Partha Dasgupta
EDUCATION Gary Thomas
EGYPTIAN MYTH Geraldine Pinch
EIGHTEENTH-CENTURY BRITAIN Paul Langford
THE ELEMENTS Philip Ball
ÉMILE ZOLA Brian Nelson
EMOTION Dylan Evans
EMPIRE Stephen Howe
ENERGY SYSTEMS Nick Jenkins
ENGELS Terrell Carver
ENGINEERING David Blockley
THE ENGLISH LANGUAGE Simon Horobin
ENGLISH LITERATURE Jonathan Bate
THE ENLIGHTENMENT John Robertson
ENTREPRENEURSHIP Paul Westhead and Mike Wright
ENVIRONMENTAL ECONOMICS Stephen Smith
ENVIRONMENTAL ETHICS Robin Attfield
ENVIRONMENTAL LAW Elizabeth Fisher
ENVIRONMENTAL POLITICS Andrew Dobson
ENZYMES Paul Engel
EPICUREANISM Catherine Wilson
EPIDEMIOLOGY Rodolfo Saracci
ETHICS Simon Blackburn
ETHNOMUSICOLOGY Timothy Rice
THE ETRUSCANS Christopher Smith
EUGENICS Philippa Levine
THE EUROPEAN UNION Simon Usherwood and John Pinder
EUROPEAN UNION LAW Anthony Arnull
EVOLUTION Brian and Deborah Charlesworth
EXISTENTIALISM Thomas Flynn
EXPLORATION Stewart A. Weaver
EXTINCTION Paul B. Wignall
THE EYE Michael Land
FAIRY TALE Marina Warner
FAMILY LAW Jonathan Herring
FASCISM Kevin Passmore
FASHION Rebecca Arnold
FEDERALISM Mark J. Rozell and Clyde Wilcox
FEMINISM Margaret Walters
FILM Michael Wood
FILM MUSIC Kathryn Kalinak
FILM NOIR James Naremore
FIRE Andrew C. Scott
THE FIRST WORLD WAR Michael Howard
FOLK MUSIC Mark Slobin
FOOD John Krebs
FORENSIC PSYCHOLOGY David Canter
FORENSIC SCIENCE Jim Fraser
FORESTS Jaboury Ghazoul
FOSSILS Keith Thomson
FOUCAULT Gary Gutting
THE FOUNDING FATHERS R. B. Bernstein
FRACTALS Kenneth Falconer
FREE SPEECH Nigel Warburton
FREE WILL Thomas Pink
FREEMASONRY Andreas Önnerfors
FRENCH LITERATURE John D. Lyons
FRENCH PHILOSOPHY Stephen Gaukroger and Knox Peden
THE FRENCH REVOLUTION William Doyle
FREUD Anthony Storr
FUNDAMENTALISM Malise Ruthven
FUNGI Nicholas P. Money
THE FUTURE Jennifer M. Gidley
GALAXIES John Gribbin
GALILEO Stillman Drake
GAME THEORY Ken Binmore
GANDHI Bhikhu Parekh
GARDEN HISTORY Gordon Campbell
GENES Jonathan Slack
GENIUS Andrew Robinson
GENOMICS John Archibald
GEOFFREY CHAUCER David Wallace
GEOGRAPHY John Matthews and David Herbert
GEOLOGY Jan Zalasiewicz
GEOPHYSICS William Lowrie
GEOPOLITICS Klaus Dodds
GEORGE BERNARD SHAW Christopher Wixson
GERMAN LITERATURE Nicholas Boyle

GERMAN PHILOSOPHY Andrew Bowie
THE GHETTO Bryan Cheyette
GLACIATION David J. A. Evans
GLOBAL CATASTROPHES Bill McGuire
GLOBAL ECONOMIC HISTORY Robert C. Allen
GLOBAL ISLAM Nile Green
GLOBALIZATION Manfred B. Steger
GOD John Bowker
GOETHE Ritchie Robertson
THE GOTHIC Nick Groom
GOVERNANCE Mark Bevir
GRAVITY Timothy Clifton
THE GREAT DEPRESSION AND THE NEW DEAL Eric Rauchway
HABERMAS James Gordon Finlayson
THE HABSBURG EMPIRE Martyn Rady
HAPPINESS Daniel M. Haybron
THE HARLEM RENAISSANCE Cheryl A. Wall
THE HEBREW BIBLE AS LITERATURE Tod Linafelt
HEGEL Peter Singer
HEIDEGGER Michael Inwood
THE HELLENISTIC AGE Peter Thonemann
HENRY JAMES Susan L. Mizruchi
HEREDITY John Waller
HERMENEUTICS Jens Zimmermann
HERODOTUS Jennifer T. Roberts
HIEROGLYPHS Penelope Wilson
HINDUISM Kim Knott
HISTORY John H. Arnold
THE HISTORY OF ASTRONOMY Michael Hoskin
THE HISTORY OF CHEMISTRY William H. Brock
THE HISTORY OF CHILDHOOD James Marten
THE HISTORY OF CINEMA Geoffrey Nowell-Smith
THE HISTORY OF LIFE Michael Benton
THE HISTORY OF MATHEMATICS Jacqueline Stedall
THE HISTORY OF MEDICINE William Bynum
THE HISTORY OF PHYSICS J. L. Heilbron
THE HISTORY OF TIME Leofranc Holford-Strevens
HIV AND AIDS Alan Whiteside
HOBBES Richard Tuck
HOLLYWOOD Peter Decherney
THE HOLY ROMAN EMPIRE Joachim Whaley
HOME Michael Allen Fox
HOMER Barbara Graziosi
HORMONES Martin Luck
HORROR Darryl Jones
HUMAN ANATOMY Leslie Klenerman
HUMAN EVOLUTION Bernard Wood
HUMAN PHYSIOLOGY Jamie A. Davies
HUMAN RIGHTS Andrew Clapham
HUMANISM Stephen Law
HUME A. J. Ayer
HUMOUR Noël Carroll
THE ICE AGE Jamie Woodward
IDENTITY Florian Coulmas
IDEOLOGY Michael Freeden
THE IMMUNE SYSTEM Paul Klenerman
INDIAN CINEMA Ashish Rajadhyaksha
INDIAN PHILOSOPHY Sue Hamilton
THE INDUSTRIAL REVOLUTION Robert C. Allen
INFECTIOUS DISEASE Marta L. Wayne and Benjamin M. Bolker
INFINITY Ian Stewart
INFORMATION Luciano Floridi
INNOVATION Mark Dodgson and David Gann
INTELLECTUAL PROPERTY Siva Vaidhyanathan
INTELLIGENCE Ian J. Deary
INTERNATIONAL LAW Vaughan Lowe
INTERNATIONAL MIGRATION Khalid Koser
INTERNATIONAL RELATIONS Christian Reus-Smit
INTERNATIONAL SECURITY Christopher S. Browning
IRAN Ali M. Ansari
ISLAM Malise Ruthven
ISLAMIC HISTORY Adam Silverstein
ISLAMIC LAW Mashood A. Baderin

ISOTOPES Rob Ellam
ITALIAN LITERATURE
Peter Hainsworth and David Robey
JESUS Richard Bauckham
JEWISH HISTORY David N. Myers
JOURNALISM Ian Hargreaves
JUDAISM Norman Solomon
JUNG Anthony Stevens
KABBALAH Joseph Dan
KAFKA Ritchie Robertson
KANT Roger Scruton
KEYNES Robert Skidelsky
KIERKEGAARD Patrick Gardiner
KNOWLEDGE Jennifer Nagel
THE KORAN Michael Cook
KOREA Michael J. Seth
LAKES Warwick F. Vincent
LANDSCAPE ARCHITECTURE
Ian H. Thompson
LANDSCAPES AND GEOMORPHOLOGY
Andrew Goudie and Heather Viles
LANGUAGES Stephen R. Anderson
LATE ANTIQUITY Gillian Clark
LAW Raymond Wacks
THE LAWS OF THERMODYNAMICS
Peter Atkins
LEADERSHIP Keith Grint
LEARNING Mark Haselgrove
LEIBNIZ Maria Rosa Antognazza
LEO TOLSTOY Liza Knapp
LIBERALISM Michael Freeden
LIGHT Ian Walmsley
LINCOLN Allen C. Guelzo
LINGUISTICS Peter Matthews
LITERARY THEORY Jonathan Culler
LOCKE John Dunn
LOGIC Graham Priest
LOVE Ronald de Sousa
MACHIAVELLI Quentin Skinner
MADNESS Andrew Scull
MAGIC Owen Davies
MAGNA CARTA Nicholas Vincent
MAGNETISM Stephen Blundell
MALTHUS Donald Winch
MAMMALS T. S. Kemp
MANAGEMENT John Hendry
MAO Delia Davin
MARINE BIOLOGY
Philip V. Mladenov
MARKETING
Kenneth Le Meunier-FitzHugh
THE MARQUIS DE SADE John Phillips
MARTIN LUTHER Scott H. Hendrix
MARTYRDOM Jolyon Mitchell
MARX Peter Singer
MATERIALS Christopher Hall
MATHEMATICAL FINANCE
Mark H. A. Davis
MATHEMATICS Timothy Gowers
MATTER Geoff Cottrell
THE MAYA Matthew Restall and Amara Solari
THE MEANING OF LIFE
Terry Eagleton
MEASUREMENT David Hand
MEDICAL ETHICS Michael Dunn and Tony Hope
MEDICAL LAW Charles Foster
MEDIEVAL BRITAIN John Gillingham and Ralph A. Griffiths
MEDIEVAL LITERATURE
Elaine Treharne
MEDIEVAL PHILOSOPHY
John Marenbon
MEMORY Jonathan K. Foster
METAPHYSICS Stephen Mumford
METHODISM William J. Abraham
THE MEXICAN REVOLUTION
Alan Knight
MICHAEL FARADAY
Frank A. J. L. James
MICROBIOLOGY Nicholas P. Money
MICROECONOMICS Avinash Dixit
MICROSCOPY Terence Allen
THE MIDDLE AGES Miri Rubin
MILITARY JUSTICE Eugene R. Fidell
MILITARY STRATEGY
Antulio J. Echevarria II
MINERALS David Vaughan
MIRACLES Yujin Nagasawa
MODERN ARCHITECTURE
Adam Sharr
MODERN ART David Cottington
MODERN BRAZIL Anthony W. Pereira
MODERN CHINA Rana Mitter
MODERN DRAMA
Kirsten E. Shepherd-Barr
MODERN FRANCE
Vanessa R. Schwartz

MODERN INDIA Craig Jeffrey
MODERN IRELAND Senia Pašeta
MODERN ITALY Anna Cento Bull
MODERN JAPAN
Christopher Goto-Jones
MODERN LATIN AMERICAN LITERATURE
Roberto González Echevarría
MODERN WAR Richard English
MODERNISM Christopher Butler
MOLECULAR BIOLOGY Aysha Divan and Janice A. Royds
MOLECULES Philip Ball
MONASTICISM Stephen J. Davis
THE MONGOLS Morris Rossabi
MONTAIGNE William M. Hamlin
MOONS David A. Rothery
MORMONISM
Richard Lyman Bushman
MOUNTAINS Martin F. Price
MUHAMMAD Jonathan A. C. Brown
MULTICULTURALISM Ali Rattansi
MULTILINGUALISM John C. Maher
MUSIC Nicholas Cook
MYTH Robert A. Segal
NAPOLEON David Bell
THE NAPOLEONIC WARS
Mike Rapport
NATIONALISM Steven Grosby
NATIVE AMERICAN LITERATURE
Sean Teuton
NAVIGATION Jim Bennett
NAZI GERMANY Jane Caplan
NELSON MANDELA Elleke Boehmer
NEOLIBERALISM Manfred B. Steger and Ravi K. Roy
NETWORKS Guido Caldarelli and Michele Catanzaro
THE NEW TESTAMENT
Luke Timothy Johnson
THE NEW TESTAMENT AS LITERATURE Kyle Keefer
NEWTON Robert Iliffe
NIELS BOHR J. L. Heilbron
NIETZSCHE Michael Tanner
NINETEENTH-CENTURY BRITAIN
Christopher Harvie and H. C. G. Matthew
THE NORMAN CONQUEST
George Garnett
NORTH AMERICAN INDIANS
Theda Perdue and Michael D. Green
NORTHERN IRELAND
Marc Mulholland
NOTHING Frank Close
NUCLEAR PHYSICS Frank Close
NUCLEAR POWER Maxwell Irvine
NUCLEAR WEAPONS
Joseph M. Siracusa
NUMBER THEORY Robin Wilson
NUMBERS Peter M. Higgins
NUTRITION David A. Bender
OBJECTIVITY Stephen Gaukroger
OCEANS Dorrik Stow
THE OLD TESTAMENT
Michael D. Coogan
THE ORCHESTRA D. Kern Holoman
ORGANIC CHEMISTRY
Graham Patrick
ORGANIZATIONS Mary Jo Hatch
ORGANIZED CRIME
Georgios A. Antonopoulos and Georgios Papanicolaou
ORTHODOX CHRISTIANITY
A. Edward Siecienski
OVID Llewelyn Morgan
PAGANISM Owen Davies
THE PALESTINIAN-ISRAELI CONFLICT Martin Bunton
PANDEMICS
Christian W. McMillen
PARTICLE PHYSICS Frank Close
PAUL E. P. Sanders
PEACE Oliver P. Richmond
PENTECOSTALISM William K. Kay
PERCEPTION Brian Rogers
THE PERIODIC TABLE Eric R. Scerri
PHILOSOPHICAL METHOD
Timothy Williamson
PHILOSOPHY Edward Craig
PHILOSOPHY IN THE ISLAMIC WORLD Peter Adamson
PHILOSOPHY OF BIOLOGY
Samir Okasha
PHILOSOPHY OF LAW
Raymond Wacks
PHILOSOPHY OF PHYSICS
David Wallace
PHILOSOPHY OF SCIENCE
Samir Okasha

PHILOSOPHY OF RELIGION
Tim Bayne
PHOTOGRAPHY Steve Edwards
PHYSICAL CHEMISTRY Peter Atkins
PHYSICS Sidney Perkowitz
PILGRIMAGE Ian Reader
PLAGUE Paul Slack
PLANETS David A. Rothery
PLANTS Timothy Walker
PLATE TECTONICS Peter Molnar
PLATO Julia Annas
POETRY Bernard O'Donoghue
POLITICAL PHILOSOPHY David Miller
POLITICS Kenneth Minogue
POPULISM Cas Mudde and
Cristóbal Rovira Kaltwasser
POSTCOLONIALISM Robert Young
POSTMODERNISM Christopher Butler
POSTSTRUCTURALISM
Catherine Belsey
POVERTY Philip N. Jefferson
PREHISTORY Chris Gosden
PRESOCRATIC PHILOSOPHY
Catherine Osborne
PRIVACY Raymond Wacks
PROBABILITY John Haigh
PROGRESSIVISM Walter Nugent
PROHIBITION W. J. Rorabaugh
PROJECTS Andrew Davies
PROTESTANTISM Mark A. Noll
PSYCHIATRY Tom Burns
PSYCHOANALYSIS Daniel Pick
PSYCHOLOGY Gillian Butler and
Freda McManus
PSYCHOLOGY OF MUSIC
Elizabeth Hellmuth Margulis
PSYCHOPATHY Essi Viding
PSYCHOTHERAPY Tom Burns and
Eva Burns-Lundgren
PUBLIC ADMINISTRATION
Stella Z. Theodoulou and Ravi K. Roy
PUBLIC HEALTH Virginia Berridge
PURITANISM Francis J. Bremer
THE QUAKERS Pink Dandelion
QUANTUM THEORY
John Polkinghorne
RACISM Ali Rattansi
RADIOACTIVITY Claudio Tuniz
RASTAFARI Ennis B. Edmonds
READING Belinda Jack
THE REAGAN REVOLUTION Gil Troy
REALITY Jan Westerhoff
RECONSTRUCTION Allen C. Guelzo
THE REFORMATION Peter Marshall
REFUGEES Gil Loescher
RELATIVITY Russell Stannard
RELIGION Thomas A. Tweed
RELIGION IN AMERICA Timothy Beal
THE RENAISSANCE Jerry Brotton
RENAISSANCE ART
Geraldine A. Johnson
RENEWABLE ENERGY Nick Jelley
REPTILES T. S. Kemp
REVOLUTIONS Jack A. Goldstone
RHETORIC Richard Toye
RISK Baruch Fischhoff and John Kadvany
RITUAL Barry Stephenson
RIVERS Nick Middleton
ROBOTICS Alan Winfield
ROCKS Jan Zalasiewicz
ROMAN BRITAIN Peter Salway
THE ROMAN EMPIRE
Christopher Kelly
THE ROMAN REPUBLIC
David M. Gwynn
ROMANTICISM Michael Ferber
ROUSSEAU Robert Wokler
RUSSELL A. C. Grayling
THE RUSSIAN ECONOMY
Richard Connolly
RUSSIAN HISTORY Geoffrey Hosking
RUSSIAN LITERATURE Catriona Kelly
THE RUSSIAN REVOLUTION
S. A. Smith
THE SAINTS Simon Yarrow
SAMURAI Michael Wert
SAVANNAS Peter A. Furley
SCEPTICISM Duncan Pritchard
SCHIZOPHRENIA Chris Frith and
Eve Johnstone
SCHOPENHAUER Christopher Janaway
SCIENCE AND RELIGION
Thomas Dixon
SCIENCE FICTION David Seed
THE SCIENTIFIC REVOLUTION
Lawrence M. Principe
SCOTLAND Rab Houston
SECULARISM Andrew Copson
SEXUAL SELECTION Marlene Zuk and
Leigh W. Simmons

SEXUALITY Véronique Mottier
SHAKESPEARE'S COMEDIES
Bart van Es
SHAKESPEARE'S SONNETS AND POEMS Jonathan F. S. Post
SHAKESPEARE'S TRAGEDIES
Stanley Wells
SIKHISM Eleanor Nesbitt
SILENT FILM Donna Kornhaber
THE SILK ROAD James A. Millward
SLANG Jonathon Green
SLEEP Steven W. Lockley and Russell G. Foster
SMELL Matthew Cobb
SOCIAL AND CULTURAL ANTHROPOLOGY
John Monaghan and Peter Just
SOCIAL PSYCHOLOGY Richard J. Crisp
SOCIAL WORK Sally Holland and Jonathan Scourfield
SOCIALISM Michael Newman
SOCIOLINGUISTICS John Edwards
SOCIOLOGY Steve Bruce
SOCRATES C. C. W. Taylor
SOFT MATTER Tom McLeish
SOUND Mike Goldsmith
SOUTHEAST ASIA James R. Rush
THE SOVIET UNION Stephen Lovell
THE SPANISH CIVIL WAR
Helen Graham
SPANISH LITERATURE Jo Labanyi
SPINOZA Roger Scruton
SPIRITUALITY Philip Sheldrake
SPORT Mike Cronin
STARS Andrew King
STATISTICS David J. Hand
STEM CELLS Jonathan Slack
STOICISM Brad Inwood
STRUCTURAL ENGINEERING
David Blockley
STUART BRITAIN John Morrill
THE SUN Philip Judge
SUPERCONDUCTIVITY
Stephen Blundell
SUPERSTITION Stuart Vyse
SYMMETRY Ian Stewart
SYNAESTHESIA Julia Simner
SYNTHETIC BIOLOGY Jamie A. Davies
SYSTEMS BIOLOGY Eberhard O. Voit
TAXATION Stephen Smith
TEETH Peter S. Ungar
TELESCOPES Geoff Cottrell
TERRORISM Charles Townshend
THEATRE Marvin Carlson
THEOLOGY David F. Ford
THINKING AND REASONING
Jonathan St B. T. Evans
THOMAS AQUINAS Fergus Kerr
THOUGHT Tim Bayne
TIBETAN BUDDHISM
Matthew T. Kapstein
TIDES David George Bowers and Emyr Martyn Roberts
TOCQUEVILLE Harvey C. Mansfield
TOPOLOGY Richard Earl
TRAGEDY Adrian Poole
TRANSLATION Matthew Reynolds
THE TREATY OF VERSAILLES
Michael S. Neiberg
TRIGONOMETRY
Glen Van Brummelen
THE TROJAN WAR Eric H. Cline
TRUST Katherine Hawley
THE TUDORS John Guy
TWENTIETH-CENTURY BRITAIN
Kenneth O. Morgan
TYPOGRAPHY Paul Luna
THE UNITED NATIONS
Jussi M. Hanhimäki
UNIVERSITIES AND COLLEGES
David Palfreyman and Paul Temple
THE U.S. CIVIL WAR Louis P. Masur
THE U.S. CONGRESS
Donald A. Ritchie
THE U.S. CONSTITUTION
David J. Bodenhamer
THE U.S. SUPREME COURT
Linda Greenhouse
UTILITARIANISM
Katarzyna de Lazari-Radek and Peter Singer
UTOPIANISM Lyman Tower Sargent
VETERINARY SCIENCE James Yeates
THE VIKINGS Julian D. Richards
THE VIRTUES Craig A. Boyd and Kevin Timpe
VIRUSES Dorothy H. Crawford
VOLCANOES Michael J. Branney and Jan Zalasiewicz
VOLTAIRE Nicholas Cronk

WAR AND RELIGION Jolyon Mitchell and Joshua Rey
WAR AND TECHNOLOGY Alex Roland
WATER John Finney
WAVES Mike Goldsmith
WEATHER Storm Dunlop
THE WELFARE STATE David Garland
WILLIAM SHAKESPEARE Stanley Wells
WITCHCRAFT Malcolm Gaskill
WITTGENSTEIN A. C. Grayling
WORK Stephen Fineman
WORLD MUSIC Philip Bohlman
THE WORLD TRADE ORGANIZATION Amrita Narlikar
WORLD WAR II Gerhard L. Weinberg
WRITING AND SCRIPT Andrew Robinson
ZIONISM Michael Stanislawski

Available soon:

PHILOSOPHY OF MIND Barbara Gail Montero
DIPLOMATIC HISTORY Joseph M. Siracusa
BLASPHEMY Yvonne Sherwood
PAKISTAN Pippa Virdee
TIME Jenann Ismael

For more information visit our website

www.oup.com/vsi/

T. S. Kemp

AMPHIBIANS

A Very Short Introduction

OXFORD
UNIVERSITY PRESS

Great Clarendon Street, Oxford, OX2 6DP,
United Kingdom

Oxford University Press is a department of the University of Oxford.
It furthers the University's objective of excellence in research, scholarship,
and education by publishing worldwide. Oxford is a registered trade mark of
Oxford University Press in the UK and in certain other countries

First edition published in 2021

Impression: 1

Published in the United States of America by Oxford University Press
198 Madison Avenue, New York, NY 10016, United States of America

British Library Cataloguing in Publication Data
Data available

Library of Congress Control Number: 2021930662

ISBN 978–0–19–884298–9

Printed in Great Britain by
Ashford Colour Press Ltd, Gosport, Hampshire

Contents

List of illustrations xvii

1 What is an amphibian? 1

2 The evolution of amphibians 25

3 Reproduction and life history 48

4 How amphibians move 72

5 How amphibians feed 82

6 The amphibians' world: sense organs and communication 90

7 Amphibians and humans 104

8 Conservation and the future of amphibians 117

Further reading 131

Index 133

List of illustrations

1 Largest and smallest anurans: (a) goliath frog; (b) Vijayan's frog **4**

(a) Frogs & Friends e.V. (b) Nyctibatrachus pulivijayani. CC-BY-SA-4.0. SD Biju, Garg S, Suyesh R, Sukesan S.

2 Urodeles: (a) fire salamander; (b) Chinese giant salamander **7**

(a) Federico.Crovetto/ Shutterstock.com. (b) © Mr. Chuan-Dong Yang.

3 A caecilian **9**

Fabio Maffei/ Shutterstock.com.

4 Section through amphibian skin **11**

5 Camouflage: (a) bird-poop frog; (b) mossy tree frog **14**

(a) Rosa Jay/ Shutterstock.com. (b) reptiles4all/ Shutterstock.com.

6 The two-phase breathing mechanism of frogs: (a) lowering the floor draws air into the buccal cavity from the nostrils and lungs; (b) raising the floor forces air into the lungs and out of the nostrils **17**

Mokele / CC BY-SA (https://creativecommons.org/licenses/by-sa/4.0).

7 The origin of tetrapods: sequence of Devonian-age stem tetrapods from the rhipidistian fish grades *Eusthenopteron*, *Panderichthys*, and *Tiktaalik* to the tetrapod grades *Acanthostega* and *Ichthyostega* **27**

8 The evolutionary relationships of the main tetrapod groups, including the genera referred to in the text **32**

9 Anthracosaurs: (a) *Pederpes*; (b) *Proterogyrinus*; (c) *Westlothiana* **34**

(a) Reprinted by permission from Springer Nature, *Nature*, An early tetrapod from 'Romer's Gap', J. A. Clack, copyright 2002.

(b) © D. Bogdanov. (c) Dorling Kindersley Ltd / Alamy Stock Photo.

10 Temnospondyls: (a) *Eryops*; (b) *Trematosaurus*; (c) the stereospondyl *Mastodonsaurus* **37**
(a) Warpaint/ Shutterstock.com. (b) © Dmitry Bogdanov. (c) Catmando/ Shutterstock.com.

11 Lepospondyls: (a) the microsaur *Pantylus*; (b) the nectridean *Diplocaulus*; (c) an aïstopod **39**
(a) © Dimitry Bogdanov. (b) Sergey Krasovskiy / Stocktrek Images / Alamy.com. (c) Illustration by Tim Morris.

12 Fossil relatives of modern amphibians: (a) the earliest anuran *Triadobatrachus*; (b) the earliest caecilian *Eocaecilia*; (c) *Amphibamus* **43**
(a) Eduardo Ascarrunz; Jean-Claude Rage; Pierre Legreneur; Michel Laurin, 'Triadobatrachus massinoti, the earliest known lissamphibian (Vertebrata: Tetrapoda) re-examined by µCT scan, and the evolution of trunk length in batrachians', Contributions to Zoology, 85 (2), (CC BY 3.0). (b) Book Worm / Alamy.com. (c) Dorling Kindersley Ltd / Alamy Stock Photo.

13 Amplexus in the red-eyed tree frog **51**
Sebastian Duda/ Shutterstock.com.

14 Courtship of the red-legged salamander. Top, the male positions himself beneath the female and deposits his spermatophore; middle, the male draws the female forwards with his tail-straddling walk; bottom, the female's cloaca is drawn over the spermatophore and picks it up **53**
After Jaeger and Schwartz 1991.

15 Amphibian larvae: (a) respiratory and feeding current of a frog tadpole; (b) surface-feeding tadpole; (c) stream-dwelling tadpole with mouth sucker; (d) larval salamander; (e) larva of the caecilian *Ichthyophis* **55–56**
(a) Lauder and Shafer 1985. (b) After Duellman and Trueb 1986. (c) After Duellman and Trueb 1986. (d) University of Chicago Press, Chicago. (e) sciencepics/ Shutterstock.com.

16 Parental care in anurans: (a) túngara frog at the nest; (b) midwife toad; (c) Darwin's frog **62**
(a) Universal Images Group North America LLC / Alamy Photo Stock. (b) COULANGES/ Shutterstock.com. (c) Bert Willaert / Alamy Photo Stock.

17 Axolotl **68**
Clement Carbillet / Biosphoto / Alamy.com.

18 Paedomorphic urodeles: (a) mudpuppy; (b) olm **70**
(a) E.R. Degginger / Alamy.com. (b) travelimages / Alamy.com.

19 (a) The salamander skeleton; (b) salamander locomotion on land showing how the

combination of waves of contraction moving down the body and placing the feet on the ground move the animal forwards **74**
(b) Emre Terim/ Shutterstock.com.

20 (a) Frog skeleton; (b) the mechanics of frog jumping showing extension of the hind legs, flight through the air, and landing on forelegs **76**
(a) makow/ Shutterstock.com. (b) spyChrome / iStock.

21 Other modes of locomotion of anurans: (a) the tree-living barred leaf frog; (b) the burrowing New Mexican spadefoot toad; (c) the aquatic African clawed frog *Xenopus*; (d) gliding frog **78**
(a) Chris Mattison / Alamy Stock Photo. (b) El Morro National Park. (c) Dewald Kirsten / Shutterstock. com. (d) Nature Picture Library / Alamy Photo Stock.

22 (a) Protrusible tongue of a typical salamander; (b) muscles of the extensible tongue of a plethodontid salamander; (c) plethodontid with tongue fully extended to catch insect **84**
(c) Stephen Deban / debanlab.org.

23 (a) The mechanism of tongue protrusion in an anuran. The stiffening muscle acts as a fulcrum, the protractor muscle flips it forwards out of the mouth, and the retractor muscle pulls it back in; (b) Protruded tongue showing sticky terminal pad **87**
(a) Gans C, Gorniak GC. 1982. Functional morphology of lingual protrusion in marine toads (Bufo marinus). American Journal of Anatomy 163: 195–222.

24 Caecilian skull, teeth, and jaw muscles **88**

25 Indian dancing frog leg-waving **94**
Nature Picture Library / Alamy Photo Stock.

26 The structure of the middle ear of an anuran **94**
After Parsons and Williams 1963. The relationships of the modern amphibians: A reevaluation. Quarterly Review of Biology 38: 26–33.

27 Marsh frog with a pair of expanded vocal sacs **97**
Małgosia Nowak-Kemp.

28 Sequence of increasingly aggressive visual poses of the red-backed salamander **100**
Jaeger and Schwartz 1991.

29 Caecilian head showing the sensory tentacle and reduced eye **102**
ephotocorp / Alamy Stock Photo.

30 *Homo diluvii testis* **107**
Historic Images / Alamy Stock Photo.

31 Cane toad consuming another frog **112**
Buiten-Beeld / Alamy.com.

Chapter 1
What is an amphibian?

Stroll along the edge of a stream or pond on a humid summer's evening and you might disturb a frog and watch it hop to safety in the water, or spot a warty-skinned toad as it crawls under the protecting vegetation. Look into the water and see an occasional slender little newt swimming amongst the waterweed in pursuit of its prey. Wander through a tropical rain forest peering carefully in the trees at little bright green tree frogs waiting to capture a passing insect, and at night listen to the cacophony of frogs seeking mates. If you turn over a large stone in North America you may be lucky enough to find a salamander hiding there, with an enormously long tongue for catching flying insects. These, the frogs and toads, the newts and salamanders, and also a far less familiar kind, the completely legless, burrowing caecilians of the tropics, make up the vertebrate class Amphibia. Among their many distinguishing features, they have moist skin through which they can both breathe and take in water. Most amphibians lay eggs in water that hatch into aquatic larvae with gills for breathing and no legs. These are the tadpoles that later metamorphose into the very different adult form.

For all their secretiveness compared to the other land vertebrate groups, over 8,000 species of amphibians have been described, and new ones are still discovered almost every year. In fact, there are almost as many of them as there are either reptiles or birds, and a lot more than the 5,500 or so species of mammals.

Amphibians are much the most abundant in those warm, tropical regions where the rainfall is good and there are plenty of marshes, rivers, and ponds. But there are also plenty that are able to live in some very inhospitable places, such as high latitudes and deserts. Famously, the wood frog can survive actual freezing of its body fluids in the sub-zero temperatures of an Alaskan winter by using special anti-freeze molecules to protect its cells. Many salamanders thrive in cool mountain streams of North America. The Australian turtle frog, so-called because its small head, short legs, and squat body remind one of a tortoise, lives in the desert. It spends most of its life in a deep, cool burrow, and only emerges after a heavy rain storm to mate and lay eggs in the short-lived pools that remain. A number of amphibians are almost completely aquatic. Amongst the frogs and toads, the African clawed toad *Xenopus* only very occasionally emerges onto land in order to migrate to a new place. Several of the salamanders are even more specialized for life permanently in water. The sirens are long, rather eel-like animals which even as adults have gills, tiny front legs, and no trace of back legs. They never come out of the water at all. The largest living amphibians, the giant salamanders, reach up to 1.8 m in length and they too spend their entire life at the bottom of pools and slow-moving rivers.

The one major habitat on Earth that is almost completely denied to amphibians is the sea. Unlike the other tetrapod vertebrates that all have waterproof skin—reptiles, birds, and mammals—an amphibian's permeable skin cannot resist the high concentration of salt. But, as with so many other generalizations about amphibians, there are exceptions. The crab-eating frog of China can spend short periods of time feeding on crustaceans in coastal rock pools. It is even able to remain indefinitely in more brackish water, an ability it shares with a handful of other species. A side effect of their intolerance to seawater is that amphibians have only rarely managed to reach and colonize remote oceanic islands.

A brief who's who of living amphibians

Zoologists divide the living amphibians into three very unequal sized Orders. The great majority are the frogs and toads, which make up the order Anura (Figure 1), a name that refers to the lack of a tail. (They are sometimes called Salientia, which comes from the Latin word for leaping, even though relatively few of them actually jump very much.) They are unmistakably recognizable by their flat, wide-mouthed head, extremely short, tailless body, and long, flexible back legs. Over 7,000 species of anurans are distributed throughout the world. Despite this large number, they are a fairly conservative group, with relatively little morphological variation. Most anurans also follow the normal amphibian lifestyle of a mainly land-living adult and an entirely aquatic, gilled larva or tadpole. This lack of variation has led amphibian taxonomists to create a large number of anuran families based on relatively small anatomical differences. In fact over fifty families are recognized, several with very few members.

The names 'frog' and 'toad' are misleading because they do not correspond to particular taxonomic groups, but are more by way of traditional folk names. The habit of distinguishing the two arose centuries ago in Britain, which has only two native anuran families. The frogs are the members of the Ranidae, smooth-skinned, good jumpers that spend a lot of time in water. They include the common frog and the edible frog. The toads, such as the common and the natterjack toads, are members of the Bufonidae. These are warty-skinned crawlers that spend most of their time on land. In other parts of the world, members of other anuran families tend to be called either frogs or toads on at best superficial features, such as skin texture or whether they are crawlers or leapers. As an example of the confusion of these names, one of the members of

1. Largest and smallest anurans: (a) goliath frog; (b) Vijayan's frog.

the highly aquatic family Pipidae, *Xenopus*, is commonly called the African clawed frog, but another, *Pipa*, is known as the Suriname toad. There is even a family, Megophryidae, commonly called the Asian toadfrogs! They include among others the smooth-skinned leaf litter frogs, and the more warty-skinned moustache toads that are so named because during the mating season the males grow horny spines along the upper lip.

Although anurans are tailless, the aptly named tailed frogs do look as if they have a short one. However, it is not actually a true tail at all, but a sort of penis, present only in the males for internal fertilization of the females. They are the members of the family Ascaphidae, which are also unique in having a backbone made up of nine instead of the usual eight vertebrae of other anurans. Because of this and other characters, ascaphids are believed to be one of the most primitive living anuran groups.

Two main families of anurans have specialized in living in trees. The Hylidae are a large group of tree frogs distributed over much of the world apart from Africa and South-East Asia, while the 400 or so species of Rhacophoridae are tree frogs confined to the latter regions. The two kinds are very similar, and a good example of the convergent evolution of adaptations for the same mode of life. They are small, only a few centimetres in length, and most of them are green in colour for camouflage. Their legs are short, and the fingers and toes end in swollen toe discs to help them grip firmly onto the branches. The bright-eyed frog of Madagascar is another similar tree frog, but is a member of a different family again, the Mantellidae. Mantellids are often highly colourful frogs, including, for example, the beautiful little golden frogs beloved of frog collectors, which are actually vivid red, orange, or yellow. Most of the mantellids other than the bright-eyed frog live on the ground or are semiaquatic. They are only found in Madagascar, and we think that their ancestor arrived by accident from Asia about 80 million years ago. Since then they have been evolving in isolation of the rest of the world, producing the several different ecological types.

A lot of anurans are burrowers, and similarly to living in trees, this is a way of life that evolved separately in more than one family, and in different parts of the world. The North American spadefoot toads, Scaphiopodidae, belong in a different family from the mainly European spadefoot toads of the family Pelobatidae. But the two share a similarly rotund body shape and a horny, spade-like extension to the hind feet which they use for digging their burrow. Their mode of life is similar too, involving remaining in or near the burrow except when they move away in search of a body of water for breeding. The Mexican burrowing toad is the only member of a third burrowing family, and it is even more specialized for subterranean life. Again it is short and squat, with a spade-like extension to its foot for digging. Its eyes are small and there is no ear drum, because neither vision nor hearing airborne sound are very important to it. It also has no teeth, and instead uses a specialized tongue to catch and eat the ants and termites it encounters during burrowing.

In contrast to these more or less permanently land-based burrowers, several other groups of anurans have adopted the opposite way of life and spend most if not all their time in water. The most prominent are the members of the family Pipidae, which includes the African clawed frog and the Suriname toad. They have a flattened body for streamlining, limbs that are long and splayed outwards, and large, webbed hind feet. Instead of calling in the usual anuran way using an air-filled vocal sac which would not work very well underwater, they communicate by a clicking sound. The huge Lake Junin frog is another fully aquatic frog, unrelated to pipids. It grows to 30 cm in length and 2 kg in weight, and spends its life in the cold water of deep lakes of the high Andes. Its enormous, 40 cm long webbed back legs are very effective for swimming. The lungs are reduced as it rarely breathes air, but to compensate, it has flaps of loose skin along its body that are permeated with blood vessels, to help it absorb enough oxygen from the water.

The second of the three amphibian orders are the salamanders and newts, called the Urodela (Figure 2) from the Greek for obvious tail, or sometimes Caudata from the Latin for tail. There are only about 700 living species, one-tenth the number of anurans. All urodeles have a slender elongated body, a long tail, and most of them have four fairly short legs. They occur on every

2. Urodeles: (a) fire salamander; (b) Chinese giant salamander.

continent except Australia and Antarctica, although the greatest number are found in North America, often in quite cool regions.

The urodeles include the largest living amphibians of all, the Japanese and the Chinese giant salamanders (Figure 2b), which reach as much as 1.5 m in length from the snout to the tail tip. At the other extreme, the Mexican pygmy salamander is no more than 4 cm long, half of which is made up of the tail.

The great majority of adult urodeles have the typical salamander form and a fully aquatic larva. Almost two-thirds of the species belong to a single family, Plethodontidae, most of which live in the Americas, although there are a few in the alpine region of Europe. The common name for plethodontids is lungless salamanders because, although being perfectly at home walking and feeding on land, they have lost their lungs and rely entirely on the skin for breathing. This is an adaptation to reduce the buoyancy of the body, making it easier for them to avoid being swept away by the fast-flowing streams of their typical upland habitat. There are also burrowing and cave-dwelling plethodontids, and even tree climbing species that use the long tail to help them hold on. Most of the European urodeles are called newts, a slightly vague term that refers to the members of the family Salamandridae. They are mostly quite small, up to around 20 cm in total body length, and they do have lungs.

There are also a number of salamanders that live more or less permanently in water. The Cryptobranchidae are the North American hellbender, and the two giant Asian salamanders. These very large amphibians spend their entire lives lying at the bottom of well-aerated streams. They are flattened and have skin folds to increase the surface area of the body for gas exchange. Food is caught by simply sucking in passing prey with their huge mouth, without having to move much, thereby saving a considerable amount of energy.

The sirens live permanently in the waters of the Mississippi basin and surrounding regions of North America. They have become even more modified for this life than the cryptobranchids by retaining some of the characters of the larva into adulthood. This way of evolving changes to the adult animal is called paedomorphosis, which means 'child-like anatomy', and includes external gills for breathing, very small front legs and no back legs at all, and neither teeth nor eyelids. At times of drought, sirens survive by burrowing into the mud. Two other urodele families have gone down the route of paedomorphosis to adapt for permanently aquatic life. One is the Proteidae, made up of the mudpuppies of eastern North America. They have four short legs and prominent red, bushy gills. The olm is a relative of the mudpuppies, but is only found in the dark, underground waters of the limestone caves around Croatia, Slovenia, and northern Italy. Olms were once believed to be baby dragons, with their pinkish colour, frilly gills, long, snake-like body, and tiny legs. As is usually the case for permanently cave-dwelling animals, their eyes are minute and only able to distinguish light from dark. But special electric sense organs on the snout help them to find their invertebrate prey. The other paedomorphic family are the amphiumas from the swamps of North America. These are almost completely without limbs, although they do have lungs instead of gills.

It has been said that the third amphibian order (Figure 3), the Caecilia (meaning hidden ones, or alternatively Apoda, without feet), are best known for being poorly known, thanks to their remote distribution and obscure habitat. About 200 species have been described, but there are undoubtedly a good many more yet

3. A caecilian.

to be discovered. They only live in tropical regions of the world, Africa, Central and South America, India, and South-East Asia. Caecilians are highly adapted for burrowing. The long, cylindrical body, the complete lack of limbs, and the very short or completely absent tail makes them look like large earthworms, an impression enhanced by the annular rings around the body that increase the purchase with the walls of the burrow. Alone amongst amphibians, some caecilians have small scales embedded in the rings to provide protection against abrasion. Also unlike anurans and urodeles, the bones of the skull are strong and firmly connected to one another so that the head is strong enough to be used to push through the mud or soil when constructing the burrow. Most species burrow in the soft soil of the forest floor, where they feed on earthworms, termites, and other insects that they come across in the burrow. A few caecilians, called typhlonectids, are aquatic and make their burrows at the bottom of a pond or stream.

Amphibian skin—the key to their success

As we shall discover in Chapter 2, the modern amphibians are a branch of the ancient group of vertebrates, called the Tetrapoda, that evolved legs and air-breathing, and started to move out of the water and onto the land. At one time, zoologists tended to see the modern amphibians as somehow failures that had never successfully completed the transition to life on dry land, unlike the reptiles. However, we now realize this was a misleading view. Certainly amphibians are very prone to drying out and overheating in the sun, and they are still dependent on water for their reproduction. But all living organisms must be well adapted to their particular habitat, otherwise they would not survive for long, and amphibians are no exception. In their case, they are uniquely well adapted to living at the interface between freshwater and land. While typical amphibians are dependent on the presence of water, the plus side to this is that, in marked contrast to reptiles, they can survive perfectly well when completely submerged in the water, protected from the heat of the sun, breathing dissolved oxygen, and using the water to protect their eggs and young. But they can also move around on land, breathing air, feeding, migrating, and so on. There are, of course, many variations of this simple account amongst different species. Some are more adapted for the aquatic side of amphibian life, others for the more terrestrial, even to the extent that a few spend their entire existence in one or the other environment. But all of them have evolved from this essentially amphibious mode of life, and the biology of living amphibians as a whole only makes sense from this point of view.

The best starting point for appreciating the life of amphibians is to look at the nature of the skin (Figure 4). It is made up of two parts. The outer part is called the epidermis, and consists of a layer of living cells that continually divide. Keratin, which is the tough and waterproof protein that makes up fingernails, birds' feathers, and reptile scales amongst other external parts of

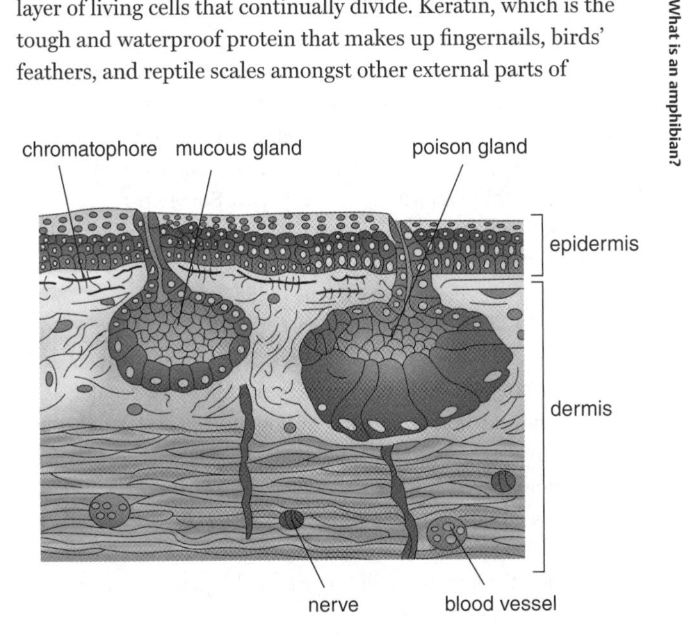

4. Section through amphibian skin.

vertebrates, is laid down as a thin layer inside the new epidermal cells. These cells then die but remain loosely attached to one another. The result is a protective layer of skin, but one that is easily penetrated between the loose dead cells by water and gas molecules. The epidermis gets damaged fairly easily, and eventually wears out due to continual abrasion from moving around on land. To compensate, it has to be renewed a few times a year. The outer layer of the epidermis is sloughed off, eaten to conserve the nutrients in it, and replaced by the new underlying layer.

The second, inner part of the skin is called the dermis, and this has a structure that is unique to amphibians. Large numbers of little goblet-shaped glands are embedded in it, each with a narrow duct opening onto the surface of the skin, so that its contents can be spread over the animal. The commonest of these are called mucous glands, and they produce a thin, watery solution of slimy mucous. The job of the mucous is to keep the skin moist so that it does not dry out in the air, which would be fatal to the living cells of the epidermis. The second kind of glands are poison or granular glands. These produce a variety of protective substances, different kinds and combinations in different species. Some are simply distasteful to predators, which soon learn to avoid that particular species. In other cases, a toxic protein is produced, which may well prove fatal to a predator that eats the individual.

The dermis also has a very rich network of fine blood capillaries running through it. These capillaries greatly increase the rate at which oxygen can be taken in and carbon dioxide expelled through the permeable skin. The ability to breathe through the skin is called cutaneous respiration, and it can take place equally well whether the amphibian is in water or on land. The bed of capillaries also lets the body take in water directly from the environment through the skin. In fact amphibians do not drink through their mouths at all.

A third kind of skin gland is found in some tree frogs, such as the orange-sided leaf frog, a bright green, orange, and white South American hylid. They are called lipid glands, and produce a fatty substance that the frog spreads over the skin using its feet, to make it temporarily waterproof. This very effectively reduces the loss of water by evaporation during the heat of the day.

No account of amphibian skin would be complete without mention of coloration. Amphibians are mostly small, highly vulnerable creatures with many predators both in water and on land. Avoiding their enemies by camouflage or cryptic colouring is therefore particularly important for them. There are cells embedded in the dermis called chromatophores, which contain coloured granules. These can be black, red, orange, or yellow. There are also cells called iridophores, so named because they contain crystalline granules that reflect white or light blue, giving the skin a shiny, iridescent appearance. Different combinations of these cells produce the wide range of colours and patterns of different species. Many frogs and toads are dull green or brown, often mottled, and this makes them inconspicuous in the low, damp vegetation and water weed where they spend much of their time. Tree frogs tend to be bright green like leaves, a hue that is produced by the interaction of yellow pigments with blue iridescence. Other tree frogs are even more remarkably camouflaged, such the Vietnamese bird-poop frog (Figure 5a), whose patchy white and black skin resembles bird droppings. The mossy tree frog (Figure 5b), also from Vietnam, looks for all the world like a patch of moss, and there are several frogs that mimic the lichen on the branch where they are resting. One of these is the North American eastern gray tree frog, which can even change its colour pattern over the course of about half an hour to match its background more exactly.

Other amphibians use colour for protection in the opposite way to crypsis, and are very brightly coloured. It is called warning, or

5. Camouflage: (a) bird-poop frog; (b) mossy tree frog.

aposematic coloration, and the conspicuousness is combined with highly distasteful or toxic skin. A predator that has experienced trying to eat such a frog or salamander soon learns to avoid others of the same memorable appearance in the future. The most spectacular examples of warning coloration are the South

American dendrobatid poison dart frogs. Amongst others, there are bright blue, vivid red, and bold yellow and black species. The skin is highly toxic, thanks to alkaloids spread from the skin glands. Curiously, the toxins are not made by the frog, but are accumulated in the poison glands from substances present in various kinds of ants and venomous centipedes that are included in the frogs' diet. The most poisonous of all species is the golden poison frog. The skin of this 5 cm long frog has enough toxin to kill ten humans. This, and a few others, are used traditionally for the points of the darts of indigenous South American hunters (although nowhere near as extensively as curare, an alkaloid found in certain plants). Amongst urodeles, a number of salamanders are similarly brightly coloured and noxious. The bright yellow patches on a black background of the fire salamander advertise its unpleasant taste, and it adds to its unpleasantness by being able to squirt a jet of toxic liquid at a threatening predator from a row of large glands down its back. Another urodele using aposematic coloration as a defence is the highly poisonous red eft, the juvenile stage of the eastern newt of North America. The European fire-bellied toads have a bright red underside which is normally hidden. But if threatened, they suddenly arch the back to flash the warning colour at a would-be predator.

Essential though it is, the skin is not the only structure involved in the amphibians' successful ability to cope with its dual life. The land presents several physiological problems not met with in the water. Atmospheric air has to be breathed; evaporation and consequent dehydration have to be avoided; and overheating when the animal no longer has the buffering effect of water has to be countered.

How amphibians breathe

Breathing in amphibians is more versatile than in any other vertebrates, because there are up to four different methods they can use. Cutaneous respiration through the skin is universal in the group, and it can be supplemented by using the lining of the mouth cavity as well, called buccopharyngeal respiration. Lungs are used for pulmonary respiration in the great majority of species, although a few kinds have lost them. Finally, branchial respiration by gills, which only works in water, is used by all the aquatic larva, and also by a small number of specialized urodeles that remain aquatic throughout life and do not lose the gills when they metamorphose into the adult stage.

Cutaneous respiration has the advantage that it can be used in water as well as on land, and that it needs very little energy. But it does have a major consequence on the animal. The rate that oxygen can be taken up depends on the area of skin exposed to the water. As a matter of simple geometry, the larger an animal is, the smaller is its surface area relative to its volume. One of the main reasons why almost all amphibians are quite small vertebrates is the need to have a relatively large surface area of skin for breathing. A few species have managed to evolve an increased relative surface area, notably the giant cryptobranchid salamanders. They have a flattened body and folds of skin that increase the area available for gas exchange. They also live at the bottom of well-oxygenated waters, and gently rock from side to side to increase the flow of water across themselves. Thanks to these adaptations, and also the fact that they are particularly slow moving, they can have a substantially greater body size than other urodeles. The African hairy frog is a more quirky example of an amphibian increasing its surface area for better cutaneous respiration. The male grows a mass of fine, hair-like extensions to its skin during the breeding season, when it is actively guarding the eggs.

The very wide mouth of amphibians provides the site for buccopharyngeal respiration. Like the external surface of the animal, the skin covering the floor and the roof of the mouth is well supplied with capillaries. Only a small proportion of the total gas exchange occurs here in most species, but it is important in

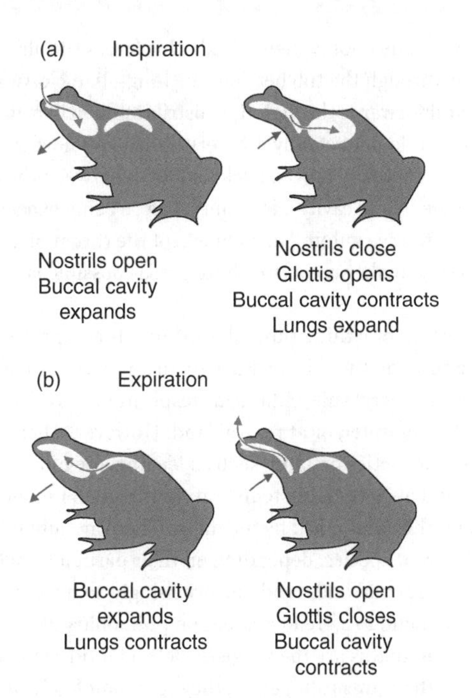

6. The two-phase breathing mechanism of frogs: (a) lowering the floor draws air into the buccal cavity from the nostrils and lungs; (b) raising the floor forces air into the lungs and out of the nostrils.

those without lungs. For example, about a quarter of the oxygen requirements of lungless plethodontid salamanders is met by buccopharyngeal respiration.

Pulmonary respiration by means of a pair of lungs can only be used when the animal is breathing air (Figure 6). The way amphibians fill their lungs is different from the method familiar to us of actively expanding a ribcage to suck the air in, because amphibians do not have ribcages.

Instead, the floor of the mouth is lowered, which sucks both fresh air through the nostrils and stale air from the lungs into the buccal cavity. The floor is then raised. This forces the air backwards through the trachea into the lungs, like blowing up a balloon, and forwards through the nostrils and out. The two streams of air do not actually mix very much, and normally the air breathed in is actually moved back and forth between the elastic lungs and the buccal cavity a few times, before being expelled. This is why we often see gulping movements of the throat of a frog, which serve to make sure as much oxygen as possible is absorbed.

How important breathing through the lungs is compared to breathing through the skin varies with an amphibian's current activity. As a general rule, cutaneous respiration provides enough oxygen when in water, or at rest on land. However, when the animal is more active on land, such as feeding, escaping, or courting, and also when the temperature is higher, more oxygen is needed and this is supplied by the lungs. There are also differences among different species, depending on their particular habitats. The more terrestrial ones tend, unsurprisingly, to have larger lungs and depend to a greater extent on pulmonary respiration. The exception to this are the lungless plethodontid salamanders. Having lost their lungs altogether, they rely completely on the skin for gas exchange, even though they spend most of their time on land. This is not a problem for them, however, because they live in upland areas, around mountain streams that keep the environment cool and humid. The more highly aquatic anurans, for example the Suriname toad, have smaller lungs and rely mainly on cutaneous respiration. But they do not lose the lungs altogether, because they are also important for making the animal more buoyant when swimming.

Gills consist of a mass of fine lamellae full of capillary blood vessels, which together have a large surface area. They are very efficient for gas exchange in water, but useless in air because the lamellae all stick together by surface tension. The gills in newly hatched larvae extend outwards into the water, and growing urodeles remain like that. Anuran tadpoles, however, soon replace these external gills with internal gills that are protected within a pouch on either side of the head. It is not entirely clear why small urodele larvae and tadpoles cannot rely on their skin alone for respiration. Probably it is because they often find themselves in warm, stagnant water that is low in oxygen, so they have to be more efficient at acquiring what little there is. At any event, the gills are lost during metamorphosis into the adult, with the exception of a few of the paedomorphic urodeles, such as the sirens, mudpuppies, and olm.

Staying moist

Amphibian skin allows oxygen to pass into and carbon dioxide out of the body, which is a great advantage for breathing as we have just seen. But the skin cannot prevent water molecules passing through equally freely, which creates a problem. Living cells can only tolerate a very small amount of water loss before they die, and they can only take in a limited amount of excess water before they burst. In dry air, amphibian skin loses water by evaporation almost as fast as if there was no barrier at all. On the other hand, when the animal is in freshwater, water molecules rapidly flow through the skin and into the living cells by osmosis. Therefore much of an amphibian's life has to be devoted to keeping its skin moist and the water level in its body within acceptable limits, although they can actually tolerate quite a high level of water loss compared to other land animals.

The most important means of keeping the water content of the body constant is simply to reduce evaporation by staying in a damp microhabitat, something that is easier for relatively small animals like amphibians to do. Even when well away from a body of water, leaf litter, ground vegetation, logs, large stones, and so on can all offer a cooler, humid atmosphere out of the sun. Most frogs and salamanders spend almost all the day in such refuges. Tree frogs find their shelter under the bark, amongst the moss covering the boughs, and tucked into the branching axils of the plants. It is mostly at night, when the conditions are more humid, that they become active, hunting prey and seeking mates. Those tree frogs that produce a waxy lipid from skin glands and wipe it all over the body spend the day squatting on a tree branch with their legs tucked up close, to reduce the exposed surface area as much as possible.

To replace the amount of water that is still inevitably lost while on land, the permeable nature of the skin is once again important, because of its remarkable ability to take up water directly. It can do this from a rainwater puddle, or even from a mere film of condensation. It is a particularly rapid and efficient way to rehydrate, because the water enters the blood capillaries directly and is immediately transported into the body, instead of having to pass through the mouth and intestine first. In many frogs and toads, the skin of the lower belly region is particularly rich in capillaries. It is called the pelvic patch, and to absorb water the hind legs are spread out backwards and the belly is pressed against the puddle or wet surface.

Burrowing is a widespread behavioural way of combating water loss, one that is taken to the extreme by anurans adapted to living in deserts, such as the spadefoot toads of southern California and Mexico. These rotund, 5–6 cm long anurans have a horny extension to the hind foot which they use to dig their burrow in sand or soil. They are able to take in up to 50 per cent of their body weight in water, by rehydrating the tissues as much as possible and filling up the large bladder. Another adaptation they have is the ability to raise the concentration of the body fluids with urea. This makes it easier for them to absorb water by the process of osmosis through the skin from only slightly damp soil. So effective at water retention are they that spadefoot toads can stay underground, remaining completely inactive for several years. They only emerge in response to heavy rainfall, when they rapidly rehydrate. As we shall see in Chapter 3, this is followed by extremely rapid mating, egg laying, and larval development.

The Australian water-holding frog is another desert burrower. When conditions are dry and hot, it fills its bladder and the cavities under the skin with water, and makes a waterproof cocoon around itself from several layers of shed skin. It can remain for as long as five years, safely buried up to a metre deep in sandy soil.

Amphibians that spend most or all of their time in water, such as the African clawed frog, the sirens amongst the urodeles, and several of the caecilians, face the opposite problem to those on land. Water continually enters the body by osmosis through the permeable skin, and salts, especially sodium chloride, are lost by diffusion out. To help combat this, the skin is somewhat less permeable to water than that of their land-based relatives, and also their body fluids are more dilute to start with. The kidneys continually remove the excess water by producing large amounts of very dilute urine, and expelling it from the bladder as often as every ten minutes or so. Meanwhile, another property of the remarkable skin of amphibians is that it can actively take up salt from the surrounding water, even though it is present at an extremely low concentration. This pumping of sodium and chloride ions across the skin against a concentration gradient is possible in freshwater, although it does take a lot of energy. However, seawater is a very different matter indeed because, far from being more dilute, it is around twice as concentrated a salt solution as amphibian body fluids. The skin is simply not capable of coping with the inflow of salt by diffusion and the outflow of water by osmosis. Almost invariably, seawater is quickly fatal to an amphibian. The crab-eating frog of South-East Asia is one of the few exceptions. It can tolerate full strength seawater for short periods, and brackish water indefinitely, and spends time in and out of rock pools or mangrove swamps feeding on crustaceans and other marine invertebrates. Technically, the way it survives is to reduce osmosis by raising the concentration of the excretory product urea and other molecules in the blood to a level that would be instantly fatal to other species.

Being the right temperature

The temperature of the body of an animal has a large effect on its activities, such as how fast it can move, digest its food, and develop and grow. There is a lower body temperature limit, below which virtually all activity ceases, called torpor, and eventually death occurs. As the body temperature increases, activity first rises and then starts to fall away as the body temperature approaches an upper limit. Above this, death due to overheating soon follows. It is very important, therefore, that the body temperature is kept as close to the optimum as much as possible, and always within the viable range. Two vertebrate groups, birds and mammals, generate their own heat and are covered in an insulating layer to keep the body temperature constantly at the optimum level. They are called endotherms, and can stay fully active over a wide range of outside temperatures. But the downside to endothermy is that it demands large amounts of energy to produce enough body heat. In practice this means collecting and eating around ten times as much food a day compared to the alternative way of regulating body temperature, called ectothermy. Amphibians, like reptiles, are ectotherms and they rely on heat from the environment, rather than generating it for themselves. Mostly they achieve this by suitable behaviour.

The moist, thin skin of amphibians prevents them from warming up by basking directly in the sun for more than a very short time. They would lose too much water by evaporation, and in any case the evaporation itself would have a counteracting cooling effect. There are exceptions to this rule though. Amphibians living at altitude and at high latitudes usually begin their day by basking for long enough to raise the temperature, before retreating into shade. The Arizona canyon tree frogs live around streams, but spend much of the day resting on the very warm boulders in the full sun, a short distance away from the water. They keep cool by evaporation, and from time to time replenish the lost water from the stream. This behaviour helps them avoid their main predator, a garter snake which forages in and alongside the streams.

Preventing overheating on a warm day on land is a matter of seeking out a cool, humid microhabitat. Burrowing species remain within their burrow, those close by ponds and streams may jump into the water, and others choose shady spots amongst the undergrowth and rocks of their habitat.

Amphibians in water, whether temporarily or permanently, cannot help being the same temperature as the water. However, water temperature does not fluctuate between day and night anywhere near as much as air temperature. Similarly, it does not vary too much between warm and cool seasons in tropical and near tropical regions.

But life in the higher latitudes of the world, where freezing temperatures occur, is a different matter. Freezing of the body fluids is fatal for the vast majority of animals, and amphibians living in such a habitat need to avoid it at all cost by hibernation. Before the onset of winter, they store a good supply of food in the liver and in special fat bodies in the abdomen. A well-protected spot is then sought. This may be deep in the leaf litter, beneath a log, in a space between the roots of a tree, or in an abandoned rodent burrow. Hibernation at the bottom of ponds is common as well, and North American salamanders, which spend most of the summer on land, frequently choose spots in fast-flowing steams that do not freeze over. At low winter temperatures, the metabolism falls to a much lower level and the food stores can last for several months if necessary. Often, however, during a warmer spell, a hibernating frog or salamander may briefly emerge to feed before returning to its haven.

Remarkably, there is a handful of amphibians that can tolerate freezing. The North American wood frog ranges all the way from Georgia into Alaska and north-eastern Canada, where it lives in forested and boggy habitats. With the onset of winter, wood frogs start to hibernate by burrowing into loose soil. A lot of the water is drawn out of the cells of the essential organs such as the heart and liver, and passed into the abdominal cavity and spaces between the muscles and the skin. When the body temperature falls below zero, this fluid starts to freeze. However, the cells have meanwhile been filled with molecules, mainly glucose, that act as antifreeze agents and prevent freezing. Survival of wood frogs down to a temperature as low as −14°C has been recorded in Alaska. The Siberian salamander can similarly survive freezing, and is the only urodele that can live above the Arctic Circle. In winter it seeks out a cavity, such as in a suitable rotting log, and here groups of up to twenty-five individuals congregate for further protection. They can survive extraordinarily low temperatures for a very long period; one specimen was reported from deep in the permafrost of Russia, at a temperature of −40°C and an estimated age of 90 years!

The converse of hibernation is called aestivation, when an animal finds a safe area to avoid dangerously high temperatures, where it remains torpid. We have already seen, for example, how desert anurans stay deep within burrows, taking a supply of water and food with them. The metabolic rate and heart rate drop dramatically, and there they remain cool until conditions above ground improve enough for them to emerge and rehydrate, feed, and mate.

The amphibians are a fairly conservative group today, not varying a great deal in their general biology. But they are the living representatives of a far greater diversity of extinct tetrapods, an evolutionary radiation that commenced with the conquest of land by vertebrates around 365 million years ago. We shall look at this history in Chapter 2.

Chapter 2
The evolution of amphibians

From our human point of view, living on dry land is the most natural thing in the world. Indeed, water is a serious danger for us. If someone falls into a river or sea who has never learned to swim, they soon drown. If they can keep themselves afloat and can breathe, they survive at most a few hours longer before fatally losing heat and succumbing to hypothermia. Even should a protective life raft be available, the level of salt in seawater is too high for the human kidney to cope with, and they die of dehydration. Compare this state of affairs with the fact that life first started in the sea 3,500 million years ago, and was completely confined to water until scarcely more than 400 million years ago. Our existence, and that of all the other kinds of terrestrial organisms, plant and animal, ought to surprise us more. The conquest of land was one of the great evolutionary revolutions of the history of life on Earth.

The origin of the tetrapods

We can begin the story of the four-legged, land-living vertebrates called tetrapods about 420 million years ago. Shortly before the start of what geologists call the Devonian Period of Earth history, a few very simple plants had established themselves along the edges of the estuaries, rivers, and lakes. Unlike its ancestors, a plant such as *Cooksonia* had evolved stronger, stiffer stems that extended a few centimetres above the body of the plant, and carried the reproductive spores at the tips of the simple branches. The wind could spread the spores much more widely, which gave these plants a considerable evolutionary advantage.

Early in the 20th century, in a village near Aberdeen in Scotland called Rhynie, the geologist William Mackie made an important fossil discovery. The geological formation in which the fossils were found is the famous Rhynie Chert, which was formed about 410 million years ago from the deposits of a hot spring along the margin of a shallow lake. Many of the organisms then living were beautifully preserved by the silica dissolved in the spring water, giving us a window on to a whole lakeside community around the start of the Devonian. There were at least seven kinds of plants, some up to 2 m high and some with extensively branched stems and small leaves. Amongst them were, for example, club mosses, a primitive kind of plant still around today. What the Rhynie Chert also shows us is that by this time a completely new habitat had arisen on the land around the wetlands. The plants provided food and protective undergrowth, and increased the oxygen level in the atmosphere through photosynthesis, which together created suitable conditions for animal life. Sure enough there are also many fossils of land invertebrates: centipedes and millipedes, spiders and mites, wingless insects and large, voracious spider relatives called trigonotarbids. Gradually, as the Devonian Period progressed, more and more kinds of plant evolved, such as conifers and horsetails. Several of these were quite large trees, up to 15 m tall and better able to live in drier regions. In time whole forests had evolved, complete with undergrowth plants, scramblers, and a tree canopy above.

But there were still no land vertebrates. Amongst the fishes living in the adjacent shallow coasts, estuaries, lakes, and rivers were a number of species of a group called the lobed-fin fish, technically rhipidistians. Fossils of one such, duly named *Eusthenopteron* (Figure 7), were discovered by the palaeontologist F. E. Whiteaves

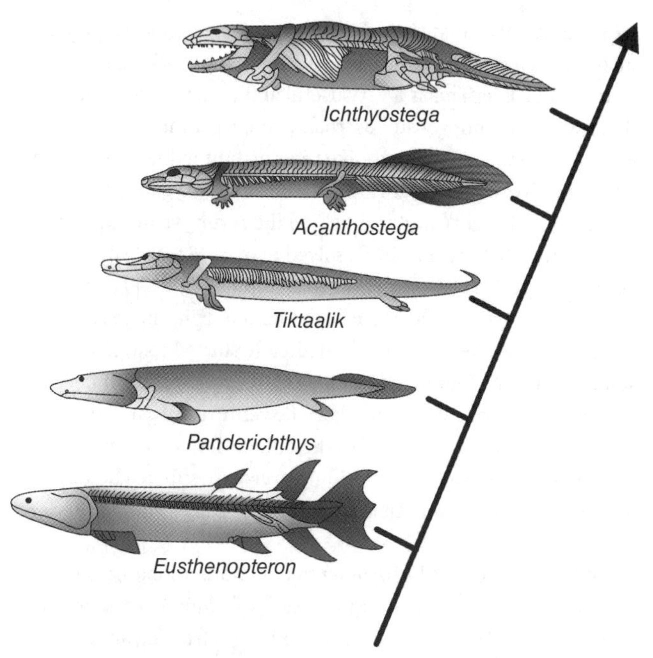

7. The origin of tetrapods: sequence of Devonian-age stem tetrapods from the rhipidistian fish grades *Eusthenopteron*, *Panderichthys*, and *Tiktaalik* to the tetrapod grades *Acanthostega* and *Ichthyostega*.

in 1881, and he recognized how important a find it was. Unlike typical fishes, *Eusthenopteron* has a bony skeleton resembling a simple limb extending part of the way into its fins. In the front fin, there is a relatively large single bone attached to the shoulder girdle that corresponds to the humerus bone of a tetrapod fore limb. At its outer end it is attached to two bones, smaller versions of the radius and ulna of a tetrapod limb, and then there are a few small bones beyond that. However, the rest of the fin is still supported by fin rays embedded in the skin, as in a typical fish fin. The smaller hind fin is similarly supported by bones comparable to those of a tetrapod hind limb, namely the femur and, beyond that, the tibia and the fibula. This internal skeleton gives the fins more strength, so that the fish can move along and steer by using them to push on the bottom, instead of having to swing the tail from side to side. This is a very useful ability for a fish living in shallow water amongst stones, rocks, and vegetation. Rhipidistians also had lungs, like their distant living relatives, the lungfish. They could breathe air as well as get oxygen from the water using fish gills, and this helped them survive in shallow, warm water that was low in dissolved oxygen. These features and other parts of the anatomy show that rhipidistians had some characters of typical fish and others found only in the tetrapods. They were intermediates that had already started to evolve the adaptations a vertebrate needs in order to live on land. Indeed, it is likely that rhipidistians could briefly venture out onto an adjacent muddy bank, although the difficulties of moving over the ground, catching food, and avoiding over-exposure to the sun would soon have driven them back into the safety of the water.

Since the discovery of *Eusthenopteron* and other lobe-finned fish, a sequence of ever more tetrapod-like fossils have been found in rocks from the Upper Devonian Period. A metre long animal called *Panderichthys* (Figure 7), found in Eastern Europe, has a rather flat body, eyes on the top of the head, and a larger fin skeleton. Discovered in Canadian rocks, *Tiktaalik* represents an even further advanced stage. It has a yet more robust skeleton in the fins, and the wrist is flexible to make it easier to manoeuvre around in shallow water and occasionally on land. It also has a longer snout, which is a feature of the tetrapod head. However, *Tiktaalik* (Figure 7) still has fin rays in the fins, large gills, and heavy fish scales covering its body.

In 1932, a Swedish palaeontologist, Gunnar Säve-Söderburgh, introduced the world to one of the most important of fossil vertebrates. *Ichthyostega* (Figure 7) was discovered in Upper Devonian rocks of East Greenland, and therefore lived at the same time as the finned rhipidistians. However it differs dramatically from them, because its large paired appendages, front and back, end in separate bony digits instead of fin vanes supported by fin rays embedded in the skin. By definition they are legs, and *Ichthyostega* was thus the first such tetrapod to be found. It has other features of the tetrapod skeleton too. For example, the eyes lie about halfway between the tip of the snout and the back of the skull. Behind the head, there are pairs of robust bony ribs attached to the vertebrae, making a rib cage suitable for supporting an animal's body as it moved around on land. On the other hand, its rhipidistian fish heritage is still evident from characters such as a long tail fin supported by fin rays, and a robust skeletal support for gills.

Acanthostega (Figure 7) is a second kind of tetrapod found in the Upper Devonian rocks of East Greenland. We know its skeleton in a lot more detail than we do *Ichthyostega*'s, thanks to the work of the Cambridge palaeontologist Jenny Clack. The most surprising discovery about *Acanthostega* was that yes, it has digits and so is technically a tetrapod, but these are quite small, and there are eight of them in each foot rather than the usual five of later tetrapods. In fact the limbs as a whole are short and paddle-like, and must have been of limited use for walking on land. Furthermore, the tail fin is large, another sign that *Acanthostega* was still an adept swimmer, whatever its walking ability might have been. The gill skeleton is as large as in rhipidistians, meaning that breathing in water was still very important, although it has perfectly good nostrils for breathing air as well. Taking all the anatomy into account, it seems as if these earliest tetrapods were still mainly, perhaps even solely aquatic animals. Maybe the tetrapod limb with digits was originally an adaptation for moving and feeding on the bed of shallow water, rather than for walking on dry land. Whether this is true or not, the evolution of limbs from fins was certainly a prerequisite for the invasion of land by vertebrates.

Palaeontologists studying the fossils have frequently wondered what the selection force was that drove the evolution of one particular group of rhipidistian fishes onto land, and they have offered numerous answers. One early suggestion was that the first tetrapods evolved limbs for moving overland, to seek new, larger bodies of water when their existing homes dried out during the annual hot season. Others believed that it was selection for feeding on the newly evolved land invertebrates; or for escaping predatory fish; or for breathing air at times when the oxygen level in warm water was low; or for warming up in the sun; the list goes on. With our growing understanding of evolution, we have now come to realize that it is too simple to think that there is just one single adaptive force driving a major evolutionary transition such as the origin of tetrapods. The sequence of ancestors and descendants, from a completely aquatic fish to a fully terrestrial tetrapod, must have been driven by selection for all aspects of the environment at the same time. Adaptations for consuming terrestrial food, for breathing air, for avoiding aquatic predators, for finding new breeding sites, for warming up during the daytime, for seeing, hearing, and smelling airborne stimuli, and so on must all have evolved together. Each character changed bit by bit and hand in hand with the rest, eventually arriving at a very different kind of organism by a process called correlated progression.

The emergence onto land was possible by this means because the change from shallow water as a habitat to the nearby land as a habitat involved a gradual change in ecological conditions. The need to use the paired appendages for movement, to breathe air instead of water, to catch and consume land-based prey, to withstand desiccation and overheating, and to use air-transmitted light, sound, and chemical stimuli had to evolve in coordination, not piecemeal. Although not many intermediate stages in this transition from fish-grade to fully evolved tetrapod have yet been discovered in the Devonian fossil record, there are enough to show how changes in different parts of the skeleton did occur at the same time, to equip the organisms for an increasingly land-based existence. The snout in front of the eyes became longer to make more room for the olfactory organ because the sense of smell works much better in air. The skull became stronger to catch prey on land by a snapping bite, instead of the suction mainly used by fish for taking in food. The branchial skeleton behind the head that supports gills in fish became smaller and was eventually lost in the adults, as the lungs took over more and more of the breathing. The rib cage gave the body support against gravity on land. The increase in the size of the bones of the paired appendages, and the replacement of fin rays by digits increased support and traction. Firmer attachment of the limbs to the shoulder and hip girdles meant larger walking forces could be applied, which increased the speed and control of locomotion. In addition to these anatomical changes recorded in the fossil record, there must have been equally important ones we know nothing directly about, because they do not show in fossils, such as an increased ability of the body to resist drying, and new reproductive habits.

The great Carboniferous radiation

The Devonian Period was succeeded by the Carboniferous Period, named after the vast deposits of coal that eventually formed from the buried plant matter. It was a time of rich vegetation along the margins of great meandering rivers and lakes, and of lush inland forests.

The level of carbon dioxide in the air was high, much higher than today, and this encouraged plant growth, and created a warm, humid environment. Over the course of about 60 million years, the Carboniferous Period witnessed a great radiation of new kinds of tetrapods (Figure 8). However, the start of this phase of life is still rather hazy. One of the most dramatic things discovered by palaeontologists about the history of life on earth is the sequence of what are called mass extinctions. These are occasions when a large number of species, over half of them, go extinct within a geologically brief length of time, only to be replaced after a

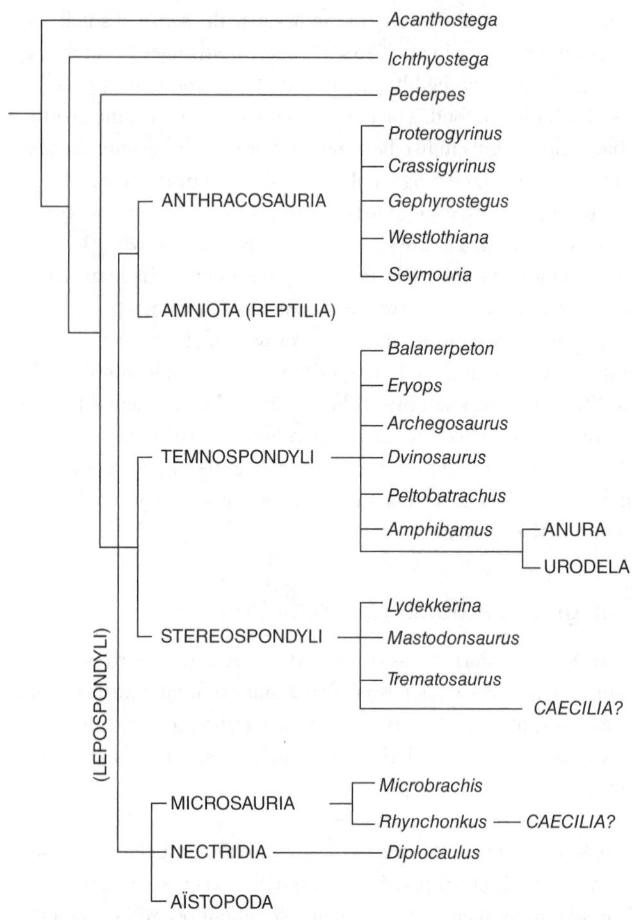

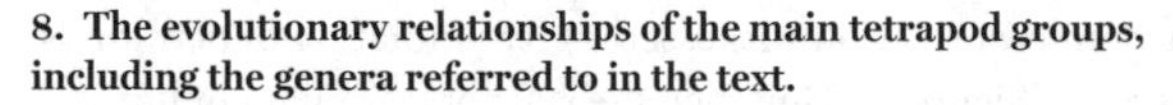
8. The evolutionary relationships of the main tetrapod groups, including the genera referred to in the text.

recovery period by the descendants of the relatively few survivors. The end of the Devonian, about 360 million years ago, is marked by just such an event. *Ichthyostega* and *Acanthostega* along with the few other known Devonian tetrapods disappeared from the fossil record, and in due course a range of new kinds of tetrapod appeared in their place. But there is a long gap of about 20 million years in between, during which very few tetrapod fossils have so far been found. It is named 'Romer's Gap' after the great American palaeontologist Al Romer, who first identified it. Whether the absence of tetrapod fossils is because the animals were particularly rare, or because they were rarely preserved as fossils for some reason, is hard to tell. But today there are strenuous efforts being made to unearth more fossils dating from this part of the Early Carboniferous in rocks of Scotland and Nova Scotia. Several fragments of various kinds of tetrapods have come to light. The most important discovery so far is a nearly complete skeleton of one named *Pederpes* (Figure 9a). Its head is less flat than the Devonian tetrapods, and the eyes face more forwards, like a land- rather than a water-dwelling creature. The limbs have five digits like most later tetrapods, and the feet point forwards rather than sideways, a feature of a truly land-walking animal. *Pederpes* was unquestionably able to live and move around successfully on land, although the feet were rather paddle shaped, indicating that they could be used for swimming as well.

Once into the later part of the Carboniferous, the roll call of fossil tetrapods increases rapidly. The warm, swampy, richly forested environment was excellent for the amphibious mode of life. Often there was little oxygen dissolved in the shallow, stagnant waters, and the organic remains of dead plants were not completely broken down but eventually turned into coal. Fortunately for palaeontologists, the remains of dead animals were occasionally buried along with the plant debris, and preserved as fossils in the coal. The huge amounts of coal mined by hand rather than mechanically during the 19th century led to the discovery of many fossil amphibians and other animals.

One of the main groups were called anthracosaurs, for example *Proterogyrinus* (Figure 9b), which has been found in coal mines of both Scotland and West Virginia in America. It is a large

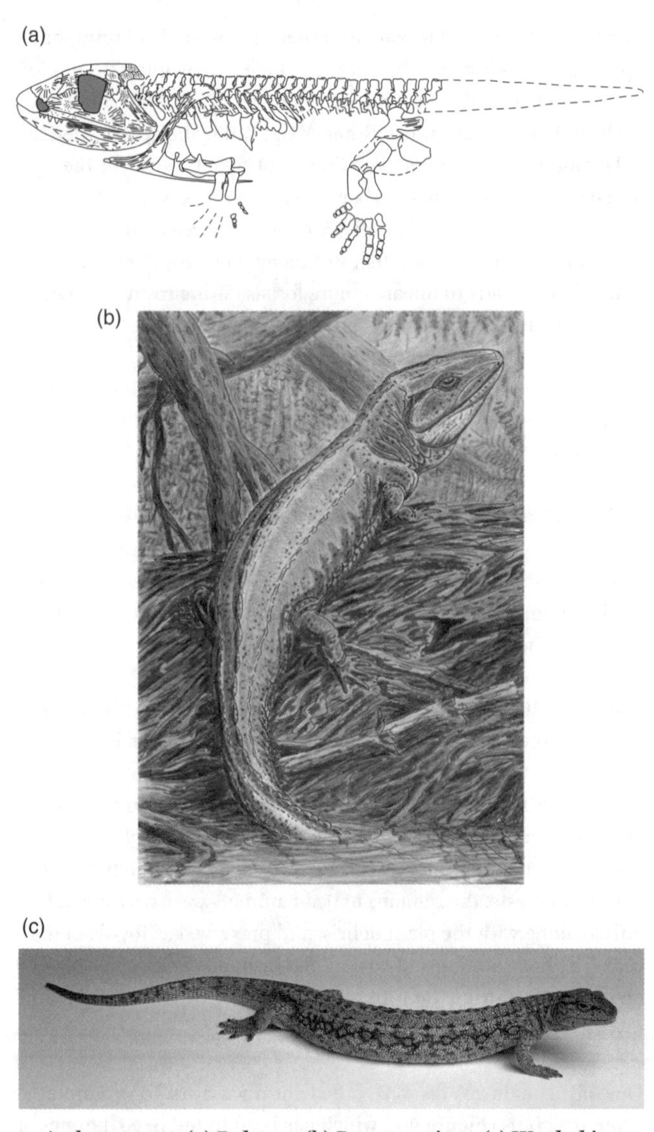

9. Anthracosaurs: (a) *Pederpes*; (b) *Proterogyrinus*; (c) *Westlothiana*.

amphibian, for the body is about 75 cm long plus a powerful tail of about the same length. The head looks rather like a crocodile, with a long snout and jaws bearing rows of small, sharp teeth suitable for catching fish, plus a few pairs of larger tusks to help hold the prey. An odd characteristic of anthracosaurs is that the cheeks of the skull were attached to the skull roof by a soft tissue hinge, so that they could swing sideways to increase the volume of the mouth. This created suction to help take in food. The four legs are a good size, and *Proterogyrinus* was able to move around effectively on land. The large tail, carrying an upper and a lower fin along its length, show that it was also a good swimmer. Other anthracosaurs differ especially in the size of the body, and how large the legs are. *Crassigyrinus* was a ferocious, aquatic predator. The body and tail together are about 2 m long, but its limbs are tiny, and could have been no more than steering organs when swimming. In contrast, *Gephyrostegus* was only around 12 cm from the tip of the snout to the anus, but it had large, strong legs and must have been a terrestrial rather than an aquatic hunter, feeding mainly on land arthropods.

In the light of the overall picture of tetrapod evolution, this is a good moment to mention the reptiles, or more exactly the Amniota, which is the name of the single group of tetrapods consisting of the reptiles, birds, and mammals. The most important feature of amniotes is that their eggs have a waterproof outer layer and large food store of yolk, so they can be laid on land and develop directly into miniature adult-like juveniles. They are called amniotic eggs, and remove the need for an aquatic larval stage. Stan Wood was the most successful recent collector of Carboniferous fossils in Scotland, and his most famous discovery is called *Westlothiana* (Figure 9c), after West Lothian, the area where it was found. When palaeontologists first studied the fossil, they believed it to be the earliest amniote ever found, which caused great excitement. With the help of new specimens, we now know that *Westlothiana* was too primitive in a number of features to be classified as a reptile, but that it is actually an anthracosaur closely related to amniotes. A much later fossil, called *Seymouria*, is a member of another group of anthracosaurs. It is particularly well adapted for life on land, with a strongly built skull and very reptile-like vertebral column and limbs. For many years an argument raged about whether *Seymouria* was an amphibian or a reptile. Eventually a small *Seymouria*-like fossil named *Discosauriscus* was discovered, as well as juvenile specimens of *Seymouria* itself. They had external gills, proving that *Seymouria* and its close relatives still had aquatic larvae rather than amniote eggs, and therefore are not amniotes. Nevertheless, *Seymouria*, like *Westlothiana*, is believed to be fairly close to the origin of the true amniotes, and few doubt that the amniotes evolved from the anthracosaur branch of early tetrapod evolution.

The temnospondyls form the next large branch of fossil amphibians. *Balenerpeton* resembles a salamander in body shape. It is about 17 cm from the tip of the snout to the base of the tail, and has four well-developed legs and a fairly short tail. The most striking feature of its head is a pair of huge spaces, called palatal fenestrae, in the roof of the mouth. This, as we shall see later, is matched in the frogs and toads amongst the living amphibians. Also like the anurans, *Balanerpeton* has a special notch at the back of its head, for an eardrum suitable for detecting airborne sound waves. It lacks the sensory organs on its head called lateral lines which amphibians that are mainly adapted for life in water have. From an ancestor like *Balanerpeton*, the temnospondyls evolved during the Carboniferous and succeeding Permian Periods into the most diverse of the amphibian groups. *Eryops* (Figure 10a), for example, is one of the most common finds in the rich fossil-bearing Permian red beds of Texas. It was a robustly built animal that reached over 2 m in length, making it one of the largest tetrapods of its time. The limbs and limb girdles supporting them were massive and *Eryops* lumbered around the marshes, sluggish rivers, and lakes hunting its food. The skull is huge and flat, and the rows of sharp teeth, supplemented by fangs inside the mouth, were used for catching fish and other smaller tetrapod prey. *Archegosaurus* (Figure 10b) is

10. Temnospondyls: (a) *Eryops*; (b) *Trematosaurus*; (c) the stereospondyl *Mastodonsaurus*.

a narrow snouted temnospondyl of which hundreds of specimens have been found in Permian rocks of Central Europe. It was much more crocodile-like in shape than *Eryops*, swimming by means of its shorter limbs and long powerful tail, and using the narrow snout to grab fish in the water.

One of the oddest temnospondyls comes from Russia. Even adult specimens of *Dvinosaurus* have a set of five pairs of large gill bars just behind the skull. Like the paedomorphic urodeles we met earlier, they suppressed complete metamorphosis and remained permanently aquatic animals. *Peltobatrachus* is another strange creature, this one from Africa. Like *Eryops*, it had a large head and well-developed limbs for walking on land. But the back and the tail were protected by large, bony scales like an armadillo.

The two main kinds of amphibians we have met so far, anthracosaurs and temnospondyls, are mostly relatively large animals, few less than half a metre or so in length. The third group to evolve in the Carboniferous are called lepospondyls, a very mixed bag of small animals. Palaeontologists are far from sure that all the lepospondyls are even closely related to one another and whether or not they have anything to do with the origin of modern amphibians.

Microsaurs are the most normal-looking lepospondyls. Some such as *Pantylus* (Figure 11a) resembled lizards, with four substantial limbs. They probably spent most of their life out of water. *Microbrachis* resembled more a salamander, for its limbs are short and its tail is flattened from side to side for swimming. The aïstopods (Figure 11c) were much stranger lepospondyls, for they had a long, snake-like body supported by about 100 vertebrae, but no trace at all of limbs or even any remnant of limb girdles. They were probably burrowers on land or in water, like the modern caecilians which they resemble in external body form. The nectridians include the bizarre lepospondyl *Diplocaulus* (Figure 11b), which has been called the 'hammerhead salamander'

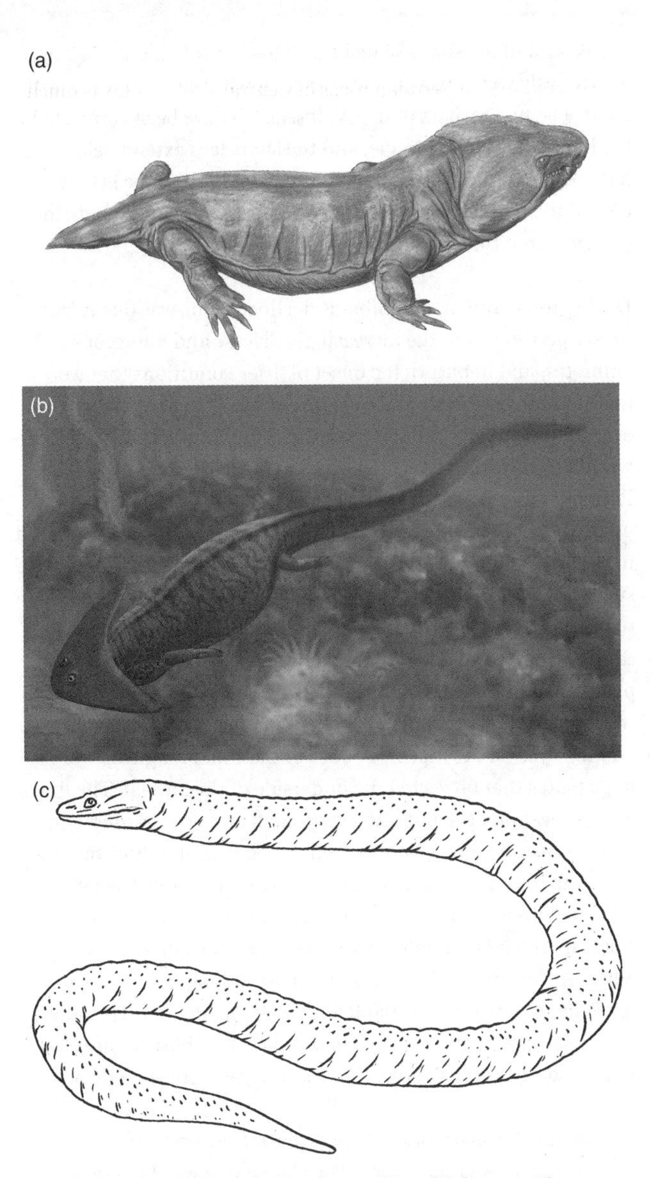

11. Lepospondyls: (a) the microsaur *Pantylus*; (b) the nectridean *Diplocaulus*; (c) an aïstopod.

on account of the shape of its head. This is a ridiculous-looking, enormously wide triangle, with small upward-facing eyes near the front. The two corners of the skull seem to have been connected to the body by a sheet of tissue, and the short legs extend below the flat body. One possible explanation for the skull shape is that it helped the animal swim upwards very fast, in pursuit of passing prey, which it then sucked into its mouth.

During the Permian, amphibians declined. This was due in part to competition from the increasingly diverse and numerous amniotes, and in part to the onset of drier conditions that were less suitable for the amphibious way of life. The final blow occurred about 250 million years ago, when the greatest of all the world's mass extinctions happened, marking the end of the Permian Period and the start of the succeeding Triassic. Over 90 per cent of the living species of animals and plants disappeared, in the sea and on the land alike. Of the amphibians, the only survivors were one or two lepospondyls, and a group of temnospondyls. The latter were called stereospondyls and were the only ones to go on to radiate quite widely during the Triassic, playing an important role as large freshwater predators. They were very flattened animals with a broad head, rounded snout, and the eyes on the top. The bones of the shoulder girdle were huge plates that protected the underside of the animal. The limbs were at best modest and often very small and weak. Lateral line grooves over the head, and the general body shape show that they lived a mainly or completely aquatic existence, using the huge head to gulp up fish. *Lydekkerina* is a small, primitive stereospondyl found quite commonly in the Lower Triassic of South Africa. Its head was only about 8 cm long, and its strong, muscular legs were well adapted for walking on land. The lateral line canals are poorly developed, compared to those of most later stereospondyls, again indicating a largely terrestrial mode of life.

In contrast, *Mastodonsaurus* (Figure 10c), a European stereospondyl, was enormous. The skull alone was 125 cm in length, and the whole body 6 m long. The limbs were small and the feet weakly ossified. Inside the mouth were several large tusks, as well as small teeth along the margins of the jaw. *Mastodonsaurus*, like several other stereospondyls of the Triassic, lay at the bottom of shallow water, waiting to suck in passing prey by quickly opening its vast mouth. *Trematosaurus* was adapted for fish eating in an entirely different way. It had abandoned the typical broad head and instead evolved a long slender snout, which it used to catch fish by actively snapping them up as it chased them through the water.

For many years palaeontologists believed that the stereospondyls had gone extinct by the end of the Triassic, about 200 million years ago. But fossils discovered in Australia prove that they were still in existence 80 million years later. There is no evidence that they survived this long in other parts of the world, but in any event it means they coexisted with early members of the modern vertebrate groups, mammals, birds, and reptiles, not to mention the modern amphibians.

The origin of the modern amphibians

The evolutionary origin of living amphibians, the anurans, urodeles, and caecilians, is one of the most exasperating topics in the study of the vertebrate evolutionary tree. One reason for the problem is that the three living groups are more different from one another than might seem at first sight. It is not even certain that they all evolved from the same common ancestor amongst the ancient amphibians. The picture has grown clearer in one way, thanks to the modern method of discovering the relationships of organisms to one another by comparing the DNA sequence of their genes. This has shown us that the living amphibians taken together are a monophyletic group, meaning that they all go back to a single common ancestor that had already separated from the common ancestor of the amniotes in the evolutionary tree.

Unfortunately, DNA is not preserved in fossils dating back millions of years, so molecular evidence is little help in discovering which of the ancient tetrapod groups are related to the modern groups. Worse still, the earliest undoubted fossils of the three modern groups do not occur until far later, tens of millions years later, than any plausible relatives amongst the Carboniferous and Permian fossil tetrapods. Such a gap allowed plenty of time for large evolutionary changes to have obscured most of the original resemblances to the ancestors.

The oldest fossil of any modern amphibian group is a very primitive frog called *Triadobatrachus* (Figure 12a). The single specimen was found over eighty years ago and it has been intensively studied by several palaeontologists ever since. It is from Early Triassic rocks of Madagascar, making it almost 250 million years old. The skull is very broad and frog-like with the characteristically huge pair of spaces in the roof of the mouth, and there is evidence that it had a frog-like tongue for feeding. Its body is shortened as in all anurans, although there are still fifteen vertebrae compared to the eight or nine of living frogs. The hind limbs are particularly intriguing, for they have some but not all the features of frog legs. The hip girdle is elongated and attached at the front to a large sacral vertebra as in frogs, but there was no movement at the joint, and there is still a row of several tail vertebrae behind it. The legs are long, but not as much as in later frogs. Some people believe that the structure of its legs shows that *Triadobatrachus* was already adapted for leaping. But others feel that the legs were specially adapted for the anuran mode of swimming in water, and that leaping only evolved later.

Despite this early existence of a primitive anuran, we have to wait no less than another 60 million years for the next fossil frog, named *Prosalirus* and discovered in Arizona. By this stage, the hind limb and pelvis were just like those of a typical living frog.

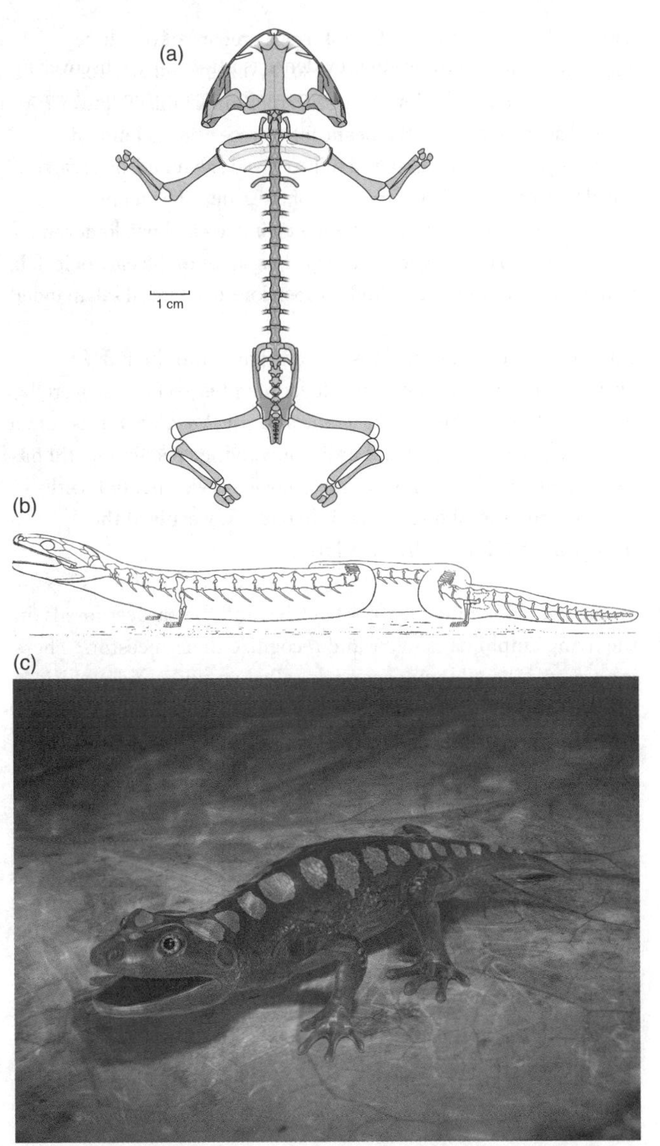

12. Fossil relatives of modern amphibians: (a) the earliest anuran *Triadobatrachus*; (b) the earliest caecilian *Eocaecilia*; (c) *Amphibamus*.

The urodeles have the least helpful early record of the three modern groups. The oldest fossils so far discovered are from 157 million-year-old, Upper Jurassic rocks in China. There are several species, such as the beautifully preserved skeleton of *Beiyanerpeton*. It was already a fully evolved salamandroid, scarcely any different from a living member of the group. The main distinction was that it had gill bars showing it was a paedomorphic species. It is 10 cm long from the tip of its snout to the base of its tail, and has the body form and limb proportions of a typical salamander.

The fossil record of caecilians extends back into the Early Jurassic of Arizona, where the oldest member so far discovered was found in the same rocks as the anuran *Prosalirus*. It is called *Eocaecilia*, (Figure 12b) and unlike any living caecilian it still has four small legs. Nevertheless, the body is very elongated, with 45–50 vertebrae, showing that it had already evolved the burrowing habit of its living relatives.

We need to look much further back beyond the earliest fossils of the living amphibians to try and recognize their ancestors. There is a group of small temnospondyls called amphibamids, which are the most likely ancient relatives of the anurans and urodeles. One example is *Amphibamus* itself (Figure 12c), which really does look somewhere between a frog and a salamander. Amongst a number of similarities to them are the very broad, flat head, large spaces in the roof of the mouth, and little teeth that have a hinge of soft tissues between the crown and the base. The vertebral column is quite short, but then so are the back legs. Amphibamids first appear in the fossil record 310 million years ago, and lived during the Late Carboniferous and Early Permian. This makes them about 50 million years older than *Triadobatrachus*, with absolutely no other fossilized intermediates yet found.

The origin of the caecilians is even less clear. The commonest opinion is that they evolved from the lepospondyls amongst the ancient amphibian groups. There is one microsaur, *Rhynchonkus*, that is particularly reminiscent of a caecilian. It has about thirty-seven cylindrical vertebrae and very short legs. Its skull bones are strong, its eyes small, and its lower jaw short, all quite similar to the earliest true caecilian, *Eocaecilia*. However there are still lots of differences between these two, and not everybody believes they are closely related. Another theory is owed to the discovery not long ago of *Chinlestegophis*. It is a very technical matter to decide exactly where it belongs in the evolutionary tree of amphibians, but seems to combine characters of the stereospondyls with a number of caecilian characters such as small and widely separated eyes, and a much smaller pair of windows in the roof of the mouth.

If this interpretation of the origin of caecilians from stereospondyls is right, then it means that all the three living amphibian groups evolved from temnospondyls, but that the caecilians branched off before the anuran and urodele lines split from each other. This is in line with the molecular evidence of the interrelationships of the modern amphibians already mentioned.

This is as about far as we can go at the moment in sorting out the divergence of the anuran, urodele, and caecilian lines from within the great Permo-Carboniferous radiation of tetrapods. Bearing in mind that there are still many millions of years with no fossil record of modern amphibian evolution, palaeontologists are sure to be faced with unexpected finds in the future that will change the picture. The very existence of the gap is intriguing. We have plenty of fossils of other kinds of early amphibians, ranging from ones living their whole life in the water to others mainly, if not solely, on the land; there are even a few marine species. Geographically they spread through equatorial, tropical, and temperate zones. From this point of view, the near absence of modern amphibian fossils for the first 100 million years of their history is most probably because they were genuinely very rare.

While their small body size certainly would not have helped fossilization, their association with water would tend to have increased the chances of individuals being preserved.

From the Jurassic and into the following Cretaceous Periods amphibians only gradually became more abundant. But then things took off. By comparing the DNA sequences of a large number of living amphibians, taxonomists have been able to discover in more detail how they are related to one another, and also to make more reliable estimates of when the evolving lines separated from one another. The results are surprising. There was evidently a large and rapid increase in the diversity of amphibians around the end of the Cretaceous. This coincided with two evolutionary events of great ecological significance for them. The first one was the rapid spread during the Late Cretaceous of the flowering plants, which created widespread broad-leaved tropical forests. The second was the increase in the diversity of insects feeding on these plants. As most amphibians live in tropical forests and eat insects, a wonderful new range of opportunities was offered them, and they duly took advantage of it. Many new adaptive kinds evolved, notably the tree frogs.

The next important event was the great mass extinction at the end of the Cretaceous that saw over 60 per cent of species disappear including, of course, all the dinosaurs (except their bird branch) and the majority of species of many other groups. Amphibians, however, survived quite well. Members of at least five groups that had arisen before the extinction survived and immediately began a radiation that, over the next 66 million years, resulted in the rich modern fauna of amphibians we have today. How frogs, salamanders, and caecilians managed to come through the mass extinction relatively unscathed compared to other groups is not at all obvious. It may have been because of their small body size, making it easier to find microhabitats in water or underground, which helped keep them safe as global temperatures temporarily soared.

Until the last two or three centuries, this success has been maintained. The recent effects of human activities on the history of amphibians, in what is often called the Anthropocene Epoch, is the subject of Chapter 8.

Chapter 3
Reproduction and life history

With their moist, water-permeable skin, the very young stages of amphibians would be at great risk on land of losing too much water from their bodies, due to their small size. On a hot day they would soon fatally dehydrate. To avoid this hazard, the life cycle of the great majority of amphibians includes a fully aquatic, juvenile stage. This is the larva, known in the anurans as the tadpole. It feeds and grows, and eventually undergoes a rapid transformation, or metamorphosis, into the adult. Typical amphibians gain another benefit from dividing the life cycle into the two separate stages. The larva is adapted to such a different environment from its parents that competition between the two for food and living space is reduced. However, the eggs and the tadpoles are very vulnerable to aquatic predators, such as fish, other amphibians, and voracious aquatic insect larvae. It is not surprising therefore to find that an impressive range of ways have evolved for improving the survival prospects of the offspring in the form of parental care. This can be as simple as laying the eggs in a safer place or guarding a nest, while in others one of the parents actually carries the eggs or tadpoles around with them. Several members of all the three main groups of amphibians that live in cool, moist habitats have managed to abandon the aquatic stage altogether. They lay their eggs on land and these hatch directly into miniature adults. At the most extreme of parental care, a few species even bear their young live.

As parental care requires active participation by one or both of the parents, it is only suitable for species living in habitats where the adults themselves are free from serious risk of dehydration. Anurans that live in deserts have no such luxury. Instead they reproduce by an accelerated version of the normal amphibian method. Spadefoot toads and the spotted toad are often found in the drier parts of the Californian desert. Here the adults take in a large store of water and burrow underground, normally for the expected ten months between annual rains, but for several years if rains do not come. Although more or less torpid for this time, they can respond to the sound of heavy rain drops during a rain storm, and this stimulates them to dig themselves out and sit in the nearest puddle, where they quickly take in water to rehydrate. On the very first night after emerging, the males begin calling to attract females, and once paired up in a temporary pool, a female sheds her eggs into the water where they are fertilized by the male. The parents pay no more attention to the eggs, but go off to feed and fill themselves with water ready for the next burrowing. Meanwhile, development of the eggs is extremely rapid. The tadpoles have an unusually broad diet of algae, insects, and filtered micro-organisms to take advantage of the relatively little food on offer, and quickly grow. Metamorphosis can take place in as little as nine days from the eggs hatching, because to survive, it has to occur before the pool dries up. The newly emerged froglets promptly dig their own burrows to escape as the pools dry up after the end of the rain storms.

Courtship and mating

Male anurans compete with cicadas and birds as nature's most prolific sound communicators. Especially in the tropics, the night-time croaking made by hundreds of courting male frogs can be almost deafening as they advertise their presence to the females by their calls. A group of simultaneously calling frogs is appropriately called a chorus, and this may consist of a single species, or two or more different species. Carefully analysed recordings show that what to us may sound like a random cacophony of sounds is more structured. Most anuran calls consist of a sequence of repeated short noises, and an individual frog will try to time his own calls to coincide with the pauses in the calling of others. This reduces the background interference, and hopefully allows a female to distinguish his call from the rest. Each anuran species produces its own characteristic sound, and in fact there are many examples of a new species of frog being first discovered in the wild by its unique call. This may be a simple whistle, a repeated note, or a continuous trilling noise. The purpose of the calling is twofold, to indicate to other males that the caller's territory is occupied and of course to attract females. As an example, the vocalization of the American bullfrog has been studied in detail. An advertisement call indicating the individual's presence is given while it is half submerged in the water. It is a short, low-pitched croak repeated every half second or so, and is timed to try and avoid interference with the calls of nearby males in the chorus. The exact sound can convey a surprising amount of information about the caller, such as his individual identity, size, and reproductive readiness, as well as his location. If another male is encountered, a characteristic chirping call is made instead, as a warning that the territory is occupied, and this can lead to physical combat between them. While still at the water's surface, they grasp one another's arms, rear up, and engage in a sort of arm-wrestling contest until one of them gives up.

Once a male anuran has succeeded in attracting a female who is ready to mate, he grasps her from behind by his forelegs. This is called amplexus (Figure 13), and is one of the most characteristic behaviours of frogs and toads. It usually takes place in the water, and the male's hold is helped by special nuptial pads that have developed on his fingers for the purpose. His forelegs are wrapped in front of the female's hind or front legs, or in some species around her head. An odd version of amplexus is found in the rain frog of southern Africa. Here the female of this burrowing species is much larger than the male, and normal amplexus would be

13. Amplexus in the red-eyed tree frog.

difficult. Instead, she exudes a sticky substance onto her skin and the male is literally glued to her back. They can even burrow together while connected in this way.

Amplexus positions the male and the female genital openings close together, and when the female sheds the eggs from her cloaca into the water, the male simultaneously ejects the sperm from his to fertilize them. By being so close to one another, there is less chance of the sperm being swept away from the eggs, and virtually no chance of the eggs being fertilized by another male. A few species have abandoned the habit of amplexus. The Bombay night frog, for instance, has a different mode of fertilization. The male just sits on top of the female, and when ready he ejects his sperm onto her. He then dismounts while the sperm runs down her back and hind legs to reach and fertilize the eggs as they emerge from her cloaca.

A small number of anurans have evolved internal fertilization. The best known of these are the tailed frogs of North America. They breed in flowing water, where the spermatozoa are at risk of washing away before successfully fertilizing the eggs. The male has a short, tail-like extension of his cloaca. He undergoes normal amplexus with a female, but instead of shedding her eggs before they are fertilized, the female retains them until the male has used his penis-like organ to introduce the sperm directly into her cloaca to fertilize them. Only then does she lay the eggs in the water.

As with almost every aspect of anuran reproductive behaviour, the duration of amplexus is very variable. It may last less than the minute needed for actual egg laying and fertilization. In many species the male and female remain in contact for a few days, while the female seeks out a suitable place to lay her eggs. The female of the Australian tortoise frog selects a mate above ground, and then, remaining attached, they burrow and spend several weeks or even months together until conditions for egg laying are suitable.

Neither urodeles nor caecilians use vocal communication in courtship behaviour. In urodeles, chemical signals called pheromones are important between the sexes. These are chemical messengers produced in special glands that convey information about individuality, aggressive intent, and sexual readiness. Visual signals and physical contact also play a role.

The red-legged salamander is a plethodontid salamander living in wooded uplands of south-eastern USA. During the breeding season, a male first recognizes a female by her pheromones. Courtship begins on land when he physically contacts her body, and rubs her head with his own. He commences a tail-straddling walk (Figure 14), a curious manoeuvre consisting of moving beneath the female until her head contacts the base of his tail, and then waving his tail from side to side induces the female to grasp it and follow him to a suitable spot. Here a method of sperm transfer unique to urodeles takes place. The male deposits a spermatophore, which is a small, compact structure made up of a

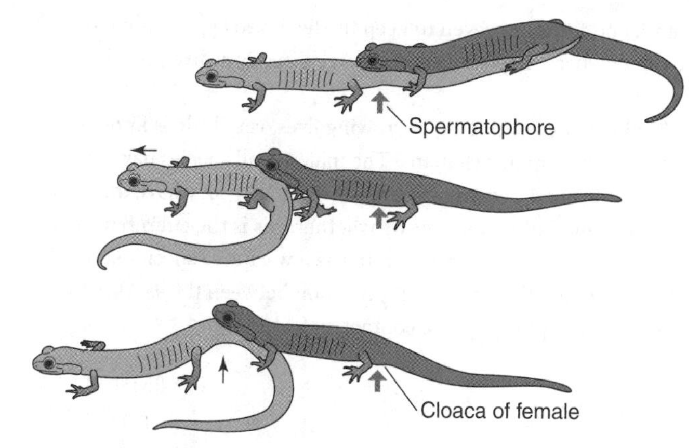

14. Courtship of the red-legged salamander. Top, the male positions himself beneath the female and deposits his spermatophore; middle, the male draws the female forwards with his tail-straddling walk; bottom, the female's cloaca is drawn over the spermatophore and picks it up.

stalk attached to the ground and a cap full of sperm. The female is drawn over it by the male, and she picks it up by the lips of her cloaca. Once inside the cloaca, the sperm swim to a special pocket called a spermotheca and there they remain. When the female ovulates, the spermotheca contracts and releases the sperm to fertilize the eggs, and she can delay this for as long as she wants. In some other species it can be a matter of months, as she awaits the right conditions for the eggs.

Zoologists have speculated on why most urodeles use spermatophores on land to transfer sperm to the female. One theory is that it gets over the danger that sperm would be washed away in flowing water if normal external fertilization took place. Another is that it helps the male partner to be sure that it is his sperm, and not that of another individual, that fertilizes his mate's eggs. One consequence of this mode of reproduction is that it makes it easier for some species of salamanders to lay the eggs on land, and for others even to keep the fertilized eggs in the mother's oviduct where they hatch and emerge directly as live juveniles.

Thanks to their secretive, burrowing lives, very little is known about courtship in caecilians. The males of all species have an extensible penis-like organ, and so we assume they have internal fertilization, but no one knows whether this is the culmination of a complicated courtship ritual. In the few captive species studied, it appears that the only communication between the sexes is by pheromones, plus physical contact, before mating.

Larval life

When we think of amphibians, we usually have in mind the various adaptations and ways of life of the adults. However, it is important not to forget that the success of a species depends just as much on how well the juvenile or larval stage copes with the hazards of life. The typical shape of an anuran tadpole is familiar from those of the frogs and toads that generations of children have watched hatching from spawn in a jam jar. The body is dark and globular, there is no sign of legs, and when first hatched there are delicate external gills for breathing in the water. These soon become protected in a gill chamber by the growth of a flap of tissue called the operculum. The tail is long and flexible, and carries a fin above and below the main axis to help swimming.

Tadpoles breathe in a similar way to fish. Water is drawn into the mouth by lowering the floor, and then pumped back into the first part of the intestine, the pharynx (Figure 15a). From here the current flows out through slits between the gills and into the gill chamber.

Finally, it leaves the gill chamber through a single hole at the back called the spiracle. On the way, oxygen is absorbed from the water by the blood vessels in the gills. The same flow of water is used for feeding, by a method called filter-feeding that is completely

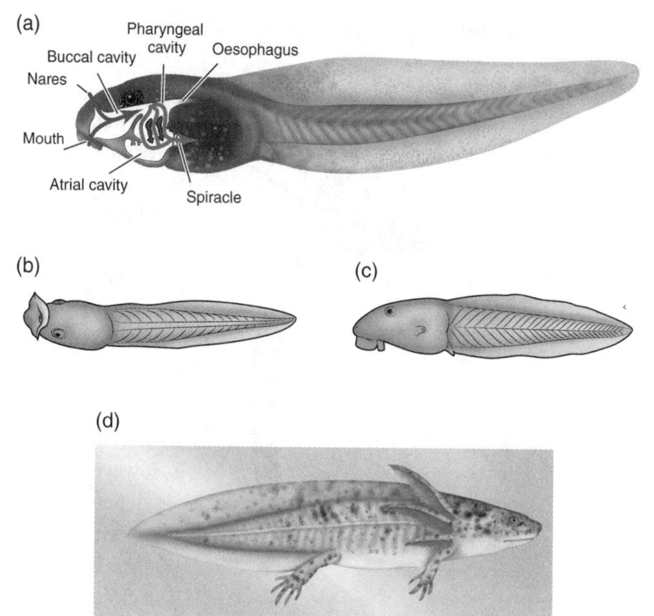

15. Amphibian larvae: (a) respiratory and feeding current of a frog tadpole; (b) surface-feeding tadpole; (c) stream-dwelling tadpole with mouth sucker; (d) larval salamander.

different from how the adults get their food. The tadpole's mouth is a wide slit, with a beak and rows of tiny teeth made of the horny skin protein, keratin. They are used to scrape algae and plant particles off the surfaces of water plants and rocks, and these are carried into the mouth by the current, along with any other fine particles suspended in the water. Before it enters the gill chamber, the food is trapped by streams of mucous in the pharynx and passed down the intestine to be digested. Plant food is less nutritious than animal food. Therefore tadpoles have a very long intestine to accommodate the amount they have to eat. It is tightly coiled up to be more compact, and hence the globular shape of the body.

(e)

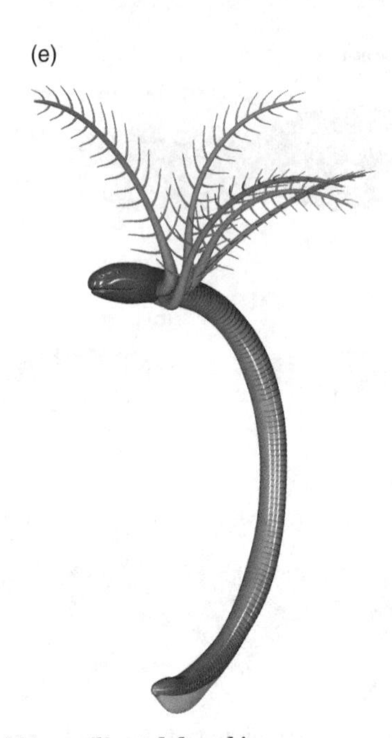

15. (e) larva of the caecilian *Ichthyophis*

Most tadpoles live near the bottom of the pond or stream, scraping up food from surfaces. But just as with the adults, there is a good deal of variation in the habits of different species. Some, like those of the phyllomedusan tree frogs inhabiting Central and South America, feed on particles floating at the surface of still or slow-moving water. Their mouth ends as an upwardly pointing funnel (Figure 15b), and the tadpole stays suspended at the surface. Others live mostly in mid-water and feed entirely by filtering out suspended food particles. Then there are species living in fast-flowing water that have evolved larger mouths to hold onto a rock. The Quito stubfoot toad tadpole has even evolved a sucker on the underside of the body, to hold on to rocks and avoid being swept away in torrential streams of its mountainous habitat in the Andes of Ecuador (Figure 15c).

The widespread South American paradoxical frog is olive green with bright green stripes and is renowned for its giant tadpoles. Those tadpoles in ponds and swamps liable to drying out are normal in size. But those living in permanent ponds and lakes grow to as much as 25 cm long, which is three to four times the length of the adult. At metamorphosis, they rapidly reduce in size, giving rise to another common name of the species, the shrinking frog.

Although tadpoles mostly eat plant food, many of them are partly or completely carnivorous. Some are even cannibals that consume other tadpoles of their own species. This happens especially in certain species that live in overcrowded conditions, for example in temporary pools in desert regions that are starting to dry up. The larger individuals start to feed on the smaller ones and so grow at a more rapid rate. They nibble the prey with their teeth, which punctures the skin and soon kills it, and then bite off and ingest small pieces. The most infamous cannibal tadpoles are those of the cane toad, the current scourge of Australia.

There is a lot less variation amongst the larvae of the urodeles, which look much more like their parents (Figure 15d). They have four legs almost from when they hatch, and of course they share with the adults a long tail. They also share a carnivorous diet, feeding on tiny planktonic invertebrates at first and later, as they grow bigger, selecting larger and larger prey. Urodele larvae keep their external gills, which are particularly large in pond-dwelling species where oxygen levels are likely to be low.

As with so much of their biology, rather little is known about the larvae of the majority of caecilians. Of those that have been investigated, only a few species have an aquatic larva, and even they hatch at quite a late stage of their development. *Ichthyophis*, for example, lays its eggs in a burrow close to water, into which the newly hatched larvae drop. The larva (Figure 15e) is worm-shaped like the adult, but has long, feathery gills and a short tail with a tail fin. Lungs soon develop, and the larvae take up an amphibious life. They tend to hide during the day amongst the vegetation, and feed at night in the water by sucking in small invertebrates.

Metamorphosis: the great crisis

The change from the larval to the adult body is called metamorphosis, and is a time of potentially great danger for an amphibian. The replacement of the larval adaptations for living in water with those of the adult for living on land means going through a short period of time in which the animal is not well suited for either. Its larval organs are being absorbed, while its adult ones are not yet fully working.

Metamorphosis is triggered by the hormone thyroxine that is produced in the thyroid gland. But the actual level of thyroxine is governed by other hormones from the brain. These are affected by the conditions of the immediate environment, such as how densely packed the larvae are, the amount of food present, and how much the pond is drying out. Therefore metamorphosis can be timed to some extent to coincide with the best circumstances. In poorer, more crowded conditions, it starts at an earlier, smaller larval stage than when there is plenty of food available in the water. Furthermore, different species of amphibians have evolved the timing of their metamorphosis to suit their habitat. The tadpoles of the desert-dwelling spadefoot toads often live in very short-lived pools that appear after rain storms. They must get to metamorphosis as quickly as possible, a matter of a couple of weeks or so. At the other extreme, the larvae of some amphibians living in areas of large, permanent bodies of water may be in a safer place than the adults on land, and can remain for several years before metamorphosis.

Metamorphosis is particularly hazardous for anurans, because the differences between the two stages of their life, tadpole and adult, are the most striking. The whole way of feeding has to change from filtering small particles out of the water to actively catching and swallowing other animals. The mouth skeleton and the horny teeth are reabsorbed, while the adult jaws develop, along with a whole new set of bones, muscles, teeth, and tongue. The long, tightly coiled intestine changes into a shorter one suitable for the new animal diet. The gills are reabsorbed and the air-breathing lungs develop. The tail is completely lost, and the hind limbs rapidly grow to their exaggerated length. The skin becomes thicker and more resistant to abrasion on land, and the sensory organs, nose, eyes, and ears, change to detect airborne signals.

Some of the adult structures actually start to develop in the tadpole before the onset of full metamorphosis. The front legs begin to grow, although they remain invisible inside the gill chamber. The hind legs first show as a pair of little buds, and then increasingly lengthen and grow toes and webbing. However, most of the dramatic change occurs very quickly, over the course of as little as a single day. The final stage is reabsorption of the tail, and the tiny froglets often start to move around on land dragging the remnants of it with them for a few days.

Metamorphosis also takes place in the urodeles, although it is much simpler. Like anurans, the gills have to be lost and the lungs develop. Changes to the head, however, are less radical because larval urodeles are already carnivorous like the adults, and require little more than the adult tongue and teeth to appear. The skin thickens, and eyelids form. Several urodeles no longer undergo metamorphosis. This is called paedomorphosis, and is a way of evolving a permanently aquatic life, something we will shortly see.

Parental care: shortening the odds

Most amphibians simply abandon their fertilized eggs in a suitable pond, lake, or slow-flowing stream and pay no more attention to them. These unprotected eggs, and the larvae which hatch from them, are vulnerable to a range of hazards. Predators include voracious insects, leeches, fish, and snakes, not to mention other amphibians, including cannibalism by adults of the same species. Bacterial and fungal pathogens are ever-present dangers, as is desiccation in many warmer parts of the world, while fast-flowing water risks them being swept away downstream. However, getting on for 20 per cent of amphibians have reduced these risks by evolving an impressive range of parental care. It may be no more complicated than the parents laying the eggs in a safer place than open water, where they would be at immediate risk of predation. Some species go further, and one or both parents remain in attendance to protect the eggs. Another method is to lay the eggs and fertilize them on land in a suitably damp place, and then actively carry the newly hatched tadpoles to water where they can feed and complete their development. Several of the most extraordinary examples consist of one or other of the parents keeping the fertilized eggs on or in part of the body, and sometimes this also includes the tadpoles.

The red-eyed tree frog of Costa Rica is the most strikingly colourful of anurans, with its red eyes, brilliant green body, blue and yellow side stripes, and orange feet. During amplexus, the eggs are laid and fertilized on the surface of a leaf directly above a pond. They are covered in a layer of mucous, which makes them stick and helps prevent them drying out. The eggs hatch after about a week, although it can occur earlier in response to the presence of a predator such as a snake. The tadpoles drop down into the water below, where they proceed to follow a normal tadpole life. Another example is the Taiwanese temple tree frog, which lays its eggs under dead leaves near to the edge of a stream.

When ready, heavy rainfall stimulates them to hatch and washes the emerging tadpoles into the water.

The grey foam-nest tree frog, which inhabits seasonally dry areas of southern Africa, has a more complicated way of using overhanging foliage. After it has rained, the female, with a male in tow, finds a suitable branch above a pool of water and covers it with a thin, mucous fluid. She lays her eggs in this, and is then joined by as many as twenty other males, who all produce sperm to fertilize the eggs, and then energetically kick the mixture into a bubbly mass of foam. The eggs develop within the nest and hatch into tadpoles that once again drop down into the water directly below.

Protection of the eggs may be provided by the parents constructing a nest floating on the surface of the water. An example of this is the little, 30 mm long túngara frog (Figure 16a) that lives in dry forests of Central and tropical South America. In the wet breeding season a pair lie together at the surface of a pool and the female produces a watery protein solution.

The male uses his hind legs to kick and churn this into the mass of foam in which the eggs are laid and fertilized. The foam nest protects the eggs from predators and bacterial infection until, after a few days, they hatch and the tadpoles swim off.

The shovel-nosed frog, or marbled snouted browser as it is sometimes called, occurs throughout the savannah of southern and central Africa. It is a burrowing species that uses its pointed snout and short, strong forelimbs to dig with, and it makes use of this ability for yet another strategy to protect the eggs before the tadpoles are deposited in water. They are laid in a burrow made near to water, and the mother protects them with a layer of unfertilized eggs and a dry, parchment-like cover. After hatching, she remains nearby, to guard the tadpoles. Once conditions are

16. Parental care in anurans: (a) túngara frog at the nest; (b) midwife toad; (c) Darwin's frog.

right, she digs a slide in wet mud between the nest and the pond for the tadpoles to wriggle down into the water.

Bromeliads are a family of tropical plants found in the Americas. Their leaves characteristically overlap at the base, and here water collects. A number of tree frogs, such as the appropriately named little bromeliad tree frog from Central America, use this receptacle as a small, safe pond to lay their eggs in, and where the tadpoles can develop. The female strawberry poison frog of Costa Rica lays her eggs in leaf litter, under a leaf, or in a small burrow, where the male fertilizes them. The male remains close by and moistens the eggs from time to time with his urine, to prevent them drying out. Once the eggs have hatched, the father summons the mother, who reappears and carries the tadpoles a few at a time on her back, high up into the tree canopy. Here she deposits them in a bromeliad pool. The tadpoles feed on small insects trapped in the pool, and the mother supplements their diet by returning from time to time to lay unfertilized eggs for them to consume. The bromeliad plants benefit from this arrangement too, by absorbing the nitrogen containing waste products of the tadpoles.

Other anuran species also feed their young directly with unfertilized eggs. Two species of jewel-eyed tree frogs were only discovered in Taiwan in 2016. They are named for their bright emerald green, or golden-yellow, eyes. The adults find a suitable water-filled hole in a tree, and there the eggs are laid and fertilized. The mother provides unfertilized eggs for the growing tadpoles to consume.

A small number of remarkable anuran species have evolved ways of protecting the eggs or tadpoles by the mother or the father carrying them around. The best-known case is the midwife toad (Figure 16b), which is widespread in Europe and North Africa, and has even been established in parts of Britain. With the arrival of the spring mating season, eggs are produced and fertilized in the water during amplexus, but then the male collects the strings of newly fertilized eggs and winds them around his hind legs. Here they are safeguarded until they hatch, when the tadpoles are deposited in a suitable pool.

The Suriname toad is almost completely aquatic. Its long, sideways extending legs and large webbed feet are inefficient for moving on land but excellent for swimming. In this species, it is the female who carries the eggs. As they are laid, and immediately fertilized by the male during amplexus, he pushes them individually onto the back of the female, where they adhere and become embedded within swollen skin. There they remain, grow, and reach metamorphosis into froglets, at which point they escape from their mother's back.

Several anurans have evolved a body pouch for carrying the eggs. The Australian pouched frog lives in the cool tropical rainforests of Queensland. The female lays her eggs in a damp spot, and when they hatch she sits on them. With help from the male, the little tadpoles wriggle through a pair of slits into a pouch between the mother's hind legs. Here they remain well protected and nourished for two or three months before metamorphosing and emerging as froglets. The marsupial frogs are a South American group, some of them living at over 3,000 m in the Andes. They have the most complete brooding behaviour of all anurans. As the female produces eggs from her cloaca, the male uses his legs to push them through a hind opening into a pouch on her back. The pouch wall swells and develops a rich blood vessel network to provide oxygen for the hatched tadpoles. These remain in the pouch, feeding on their yolk reserves for up to three or four months, before being shed into water to complete their development and undergo metamorphosis.

Evolving a pouch under the skin may be the obvious way to protect the eggs, but there are two anurans which most unexpectedly make use of a pre-existing body space as an incubation chamber. Darwin's frog (Figure 16c) is named in honour of Charles Darwin, who discovered it in Chile during his famous voyage in HMS *Beagle*. It is a small, green and brown camouflaged frog with a sharp pointed snout that would hardly be noticed if it were not for its extraordinary breeding habit. The female lays about forty eggs which the male fertilizes and guards. Shortly before they hatch, he takes them into his mouth and pushes up to about half of them into his vocal sac. There the newly hatched tadpoles grow by feeding on their yolk, and also by secretions from the wall of the vocal sac, all the way to metamorphosis, only after which they emerge as froglets.

The gastric brooding frog was discovered in the 1970s in the tropical rainforests of Queensland. Sadly though, it was already extinct by the mid-1980s. Fortunately, it had been studied for long enough to discover its astonishing method of incubating the eggs and caring for the tadpoles. After the eggs were laid and fertilized, the female swallowed them into her stomach. The jelly around the twenty or so eggs, and later the tadpoles, produced a hormone called prostaglandin that stops the stomach wall from producing hydrochloric acid. For the following six weeks, the tadpoles grew using their yolk reserves, and the mother's stomach enlarged to almost fill her abdomen. Finally, when her offspring had completed their development, she literally vomited them out as small froglets.

Viviparity: the ultimate in parental care

A small number of species in all three orders have evolved the most complete mode of parental care of all, viviparity, which is bearing the young live. The golden coqui is a species of robber or rain frog that was discovered in 1976 in Costa Rica but is already extinct. However, as in the case of the gastric brooding frog, its mode of reproduction had been studied while it was still around. The male fertilized the eggs by introducing his sperm directly into the female's oviduct, where she retained them. The mother did not provide any extra nourishment for the young, but the eggs were provisioned with enough yolk for them to complete their development. After about a month, the mother gave birth to three to six already metamorphosed froglets, capable of fending for themselves. The only other viviparous anurans we know of are the little Tournier's forest toad and its relatives in East Africa.

The strikingly marked yellow and black fire salamander is the best-known viviparous urodele. As in most urodeles, the male deposits his sperm on the ground enclosed in a gelatinous spermatophore, and then helps the female pick it up using the edges of her cloaca. However, in this species the sperm passes all the way into her oviduct to fertilize the eggs, and there they remain instead of being shed into water. After they have hatched, the larvae are nourished by the yolk that was stored in the egg, but once this is used up some of the young feed by cannibalizing others to continue growing. Typically they remain in the oviduct for a few months, including the winter hibernation period. In some individuals, the offspring are born as larvae which need to be deposited in the water to complete their development. In other individuals, metamorphosis takes place in the oviduct and the young emerge as miniature adults already able to survive on land.

Viviparity occurs in about three-quarters of the caecilians, where it has evolved at least four separate times. Unlike the viviparous anurans and urodeles, caecilian mothers do nourish their developing young from cells of the oviduct. In fact they go even further in providing for their young, because the newly emerged individuals have little teeth which they use to feed on the mother's fat-rich skin for a time.

Avoiding the water: direct development

Some amphibians avoid entirely the hazardous larval stage by laying their eggs on land, out of which miniature adults hatch, instead of gilled, legless larvae. This is called direct development, and it requires rather particular environmental conditions, because both the eggs and the tiny hatchlings are vulnerable to drying out. Some of the plethodontid salamanders of North America living close by fast-flowing upland streams have direct development. In the cool, humid conditions of their habitat, desiccation is not a significant problem, and they have the advantage that they do not have to have eggs and larvae exposed to the risk of being swept downstream by the water. The eggs of these species are specially large, and contain enough yolk for the complete development of the embryo without any need to feed.

The Taita Hills caecilian is a land-burrowing species living in the area of that name in Kenya. It lays its eggs in a nest that is guarded by the female, and they hatch directly into miniature adults. Like the live-bearing species of caecilians, these juveniles also use their teeth to nibble and consume the swollen, lipid-filled skin cells of the mother.

Direct development is rare in the anurans, although there are about twenty species whose eggs hatch directly into juvenile frogs, without undergoing metamorphosis. Most of the tiny South and Central American frogs called rain or robber frogs leave their fertilized eggs in damp habitats on land, where they hatch directly as miniature adults.

Avoiding the land: paedomorphosis

The axolotl (Figure 17), known locally as the Mexican walking fish, is a strange looking, 30 cm long creature whose home is in lakes around Mexico City. It is a sexually mature, breeding urodele but it also has several characteristics of a larva. There are three pairs of external gills, neither teeth nor eyelids, thin skin, and very short legs. In 1883, several axolotls were sent to Auguste Dumeril, a French zoologist working in Paris. To his surprise, when the aquarium they were in arrived, it contained salamanders very similar to the tiger salamanders, also of Mexico. Paedomorphosis, otherwise called neoteny, was revealed. The axolotl is a

17. Axolotl.

salamander that in its natural state remains in the larval form, continues to live in water, grows in size, and becomes sexually mature. However, if it is stressed enough, such as being transported as Dumeril's specimens were or, as we have since discovered, it is subjected to experimental conditions in the laboratory, an axolotl will undergo metamorphosis and becomes a typical adult salamander. There are quite a lot of salamander species that normally remain paedomorphic. But there are others, called facultative paedomorphs. Tiger salamanders are common throughout North America, and they can metamorphose or not depending on the local conditions. If it is dry and there is not much food in the water, metamorphosing and coming onto land is the best choice. But if there are plenty of streams and ponds with abundant aquatic food, it pays to remain larval, and continue to grow and reach maturity in the water.

Paedomorphosis has had a surprising outcome during the course of the evolution of urodeles, because in no less than four different groups it has become permanent. They are called obligatory paedomorphs, because they have no choice but to keep their main larval adaptations for a life spent entirely in water. Mudpuppies (Figure 18a) live amongst the waterweed of ponds, rivers, and lakes in the Mississippi and surrounding regions of the USA. They are up to 40 cm long, with short legs used for walking along the bottom as they hunt their prey. This consists of practically any animal, vertebrate or invertebrate, that is small enough to be sucked into the mouth. Feathery gills on each side are important for respiration. The olm (Figure 18b) is a relative of the mudpuppy, but lives exclusively in the cold waters of the underground limestone caves in the Balkan Mountains of Europe. It is very slender, and has even smaller legs than the mudpuppies. Like other cave-dwelling animals, its eyes are tiny, and are sensitive to nothing more than light and dark.

Two other kinds of obligatory paedomorphs also live in the Mississippi region. The sirens look similar to the mudpuppies,

18. Paedomorphic urodeles: (a) mudpuppy; (b) olm.

apart from having no sign of back legs. They do have lungs and can burrow into the mud if the pond or stream dries up, remaining there encased in a cocoon until the water returns. The amphiumas are known locally as congo eels. They have four minute legs, but in their case the gills are lost and the lungs are large so that they can breathe air at the surface of the water. The final group of paedomorphs are the cryptobranchids, the giant salamanders of Asia (Figure 2b) and the hellbender of North America.

Paedomorphosis is quite rare amongst animals, and why urodeles alone amongst amphibians have evolved it several times is not at all clear.

Chapter 4
How amphibians move

The three kinds of living amphibians, anurans, urodeles, and caecilians, share the same basic biology and life history, but the anatomy of the skeleton and muscles is very different amongst them. This reflects the very different ways that the locomotion of the three respective ancestors evolved. The urodeles retained the most primitive method, with a long body and tail. The anurans became far more modified by shortening the body, losing the tail altogether, and elongating the back legs. The caecilians evolved a limbless burrowing mode.

Salamanders and tadpoles—the ancestral mode

The vast majority of living fishes move through the water by undulating the body and tail fin from side to side, which creates a swimming force. As we have noted earlier, the rhipidistian fishes from which the tetrapods evolved around 380 million years ago developed bones that stiffened and strengthened the front and back pairs of fins, the forerunners of the leg skeleton of the tetrapods. They were used to gain purchase on the bottom of a shallow lake or river as the body undulated, helping drive the body forwards amongst the stones, mud, and plants. When the ancestors of the tetrapods finally started to move around on land, the limbs enlarged and began to be able to support the animal's weight off the ground. However, the basic mode of locomotion of all the groups of early fossil tetrapods still involved lateral undulation of the vertebral column, as indeed is true of lizards and crocodiles to this day.

The salamanders and newts have carried on using this ancestral mechanism of lateral undulation when swimming, but they also coupled it with limbs for walking on land. The long, slender body and pointed head are well streamlined for moving efficiently through water. There are up to 60 trunk vertebrae, which make the vertebral column very flexible between the front and back legs, while the tail is over 100 vertebrae long in some species (Figure 19a). When a salamander is in the water, a sideways contraction of the vertebral column begins at the front and passes backwards along one side of the body and tail (Figure 19b). The moving wave of contraction presses backwards against the water, and this creates the force that drives the body forwards. A wave of contraction of the other side of the body then follows, and so on. The legs, meanwhile, are held along the sides of the body so that they cause as little resistance to the forward movement of the body as possible. When on land, a salamander can still progress by lateral undulation alone, just like a snake, and they sometimes do so if suddenly startled. But usually the legs are brought into play, and serve a number of purposes. They support the animal off the ground to prevent damage to its belly, helped by a special large muscle, called the abdominal muscle that runs between the front and the hind limb girdles. When the abdominal muscle contracts, it keeps the belly from sagging downwards. The legs also provide firm anchor points onto the ground to prevent back sliding. Finally, they add to the locomotion by actively pushing backwards against the ground, driving the body forwards. Urodele walking is normally quite slow, although this is more due to the low energy available to amphibians than to the design of the body. Nevertheless, some salamanders can briefly achieve a short burst of as much as 25 kmph in threatening circumstances.

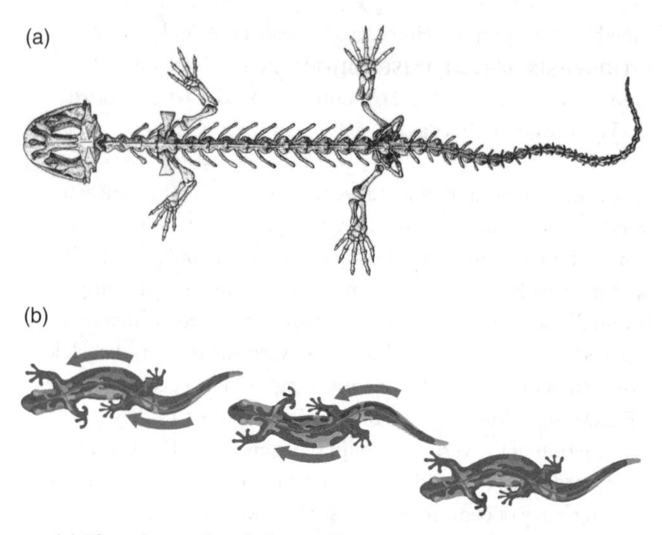

19. (a) The salamander skeleton; (b) salamander locomotion on land showing how the combination of waves of contraction moving down the body and placing the feet on the ground move the animal forwards.

The paedomorphic urodeles that fail to develop full-sized legs depend entirely on lateral undulation, although the tiny legs can still be useful. The olm, for example, uses them to help it manoeuvre between rocks and obstructions in its underground stream habitat.

The larvae of amphibians use lateral undulation for swimming, and those of urodeles behave much the same as do the adults. Anuran tadpoles also use lateral undulation, although in their case the fat, rotund body filled with the intestines means that the long tail is exclusively the motor of locomotion. It carries a fin along both the top and the bottom to increase the surface area, and incidentally it is also an important area for cutaneous respiration.

Frogs—six ways to use long legs

We usually think of the huge back legs of the anurans as adapted for leaping, and indeed they are. But they can also be used for several other modes of moving, both in water and on land depending on the species. In fact many do not leap at all. There are swimmers, hoppers and crawlers over the ground, burrowers into the ground, tree climbers, and even gliding frogs that use their feet as parachutes.

Despite this variety of modes of locomotion, the frog skeleton (Figure 20a) was originally adapted for jumping. The broad head is firmly attached to the vertebral column by a pair of contacts, called occipital condyles, either side of the opening for the spinal cord, and there is no flexible neck, which lessens the danger that the shock of landing might damage the brain. The spinal column is made up of only eight, or in a few species nine vertebrae, which is the smallest number by a long way found in any vertebrates. The column is rigid, and has lost all trace of any lateral undulation. The way to leap effectively is to have legs that can be very rapidly and greatly extended. This accelerates the body to a high enough speed that it leaves the ground and is projected through the air. The extremely long hind legs of a frog, and the way they are attached achieve this amazingly well (Figure 20b). Every bone in the leg is elongated, and every joint between them can extend through a wide angle. The thigh bone, or femur is about as long as the whole vertebral column. It is hinged at the knee to the shin bone, which is about the same length as the femur. In fact this consists of two bones, the tibia and fibula, which are separate in most tetrapods but have fused together in anurans to make the limb stronger. The main ankle bones are called the astragalus and the calcaneum. In other tetrapods these are compact and contact the ground, but in the anurans they are elongated, which adds another extensible segment to the leg. Finally, the five toe bones are also very long. The head of the

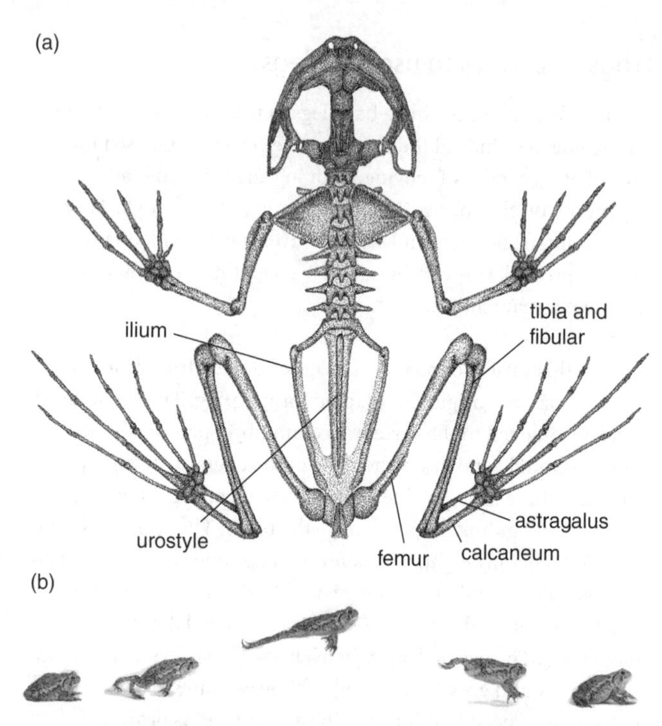

20. (a) Frog skeleton; (b) the mechanics of frog jumping showing extension of the hind legs, flight through the air, and landing on forelegs.

femur fits into the socket of the hip girdle in a fairly normal way, but the way the hip girdle attaches to the body is unique. The upper part of the hip bone is called the ilium, and like the limb bones it is also elongated. Its front tip attaches to the last vertebra, the sacral vertebra, by a mobile joint. The spinal column is continued backwards as a single, slender bone, called the urostyle, which runs between the two long ilia. When a frog is sitting on the ground at rest, its back legs are fully flexed and there is an angle between the ilia and the spinal column. When it jumps, its limb muscles rapidly extend the knee, ankle, and foot joints. Other muscles pull the femur backwards, and straighten out the angle between the ilia and the vertebral column. All these movements add together to drive the body forwards and upwards fast enough for it to be projected right into the air, often for an astonishing distance. In addition to the take off, there must also be suitable adaptations to prevent damage due to the shock of the landing. This is the role of the forelimb and shoulder girdle. The limb is relatively stout, and the broad shoulder girdle wraps around the top of the vertebral column, to which it is attached by strong muscles and ligaments. The front legs are extended forwards during the leap, and on landing, they act like strong springs that can absorb the shock.

The prodigious capacity for leaping is to be seen in the frog-jumping competitions held at various places around the world. The record single jump is held by a South African sharp-nosed frog, which is only about 6 cm long, but managed to clear 5.35 m. Another achieved over 10 m for a triple jump. American bull frogs regularly jump over 2 m.

While the frog skeleton is well adapted for leaping, especially as a way of evading predators or returning quickly to the safety of water, it has proved much more versatile. Highly aquatic species, such as *Xenopus*, the African clawed frog, have large, webbed feet (Figure 21c). They kick these backwards together to create a powerful thrust, and the body passively glides forwards before drawing the legs forwards ready for the next kick. Another, unexpected adaptation to improve the swimming in this and related pipid toads is that they can literally shorten and lengthen the body. They do this by means of a modified joint between the sacral vertebra and the ilia which lets the ilia slide forwards, shortening the body by about 20 per cent. Active extension at the joint adds to the overall length of the kick, similar to the movement of the seat of an oarsman in a racing boat. Like leaping on land, swimming in this way consists of very rapid acceleration

21. Other modes of locomotion of anurans: (a) the tree-living barred leaf frog; (b) the burrowing New Mexican spadefoot toad; (c) the aquatic African clawed frog *Xenopus*; (d) gliding frog.

of the body, which is important in avoiding predators such as pike or herons. It also has the advantage that the head does not move from side to side as the body undulates, as a fish's or a salamander's head does, which helps them capture prey that is swimming in the water directly ahead.

All anurans, even the best of leapers, can use four-legged movement on land. Sometimes they hop, which is really just short jumps using the two hind legs together and the front legs for support. At other times they crawl, using all four legs in sequence, rather like an ambling mammal. A study of Fowler's toad, a species from eastern North America, showed how it used different modes for different purposes. Crawling is slower and therefore more suitable for seeking out worms, slugs, and insects in the undergrowth. But hopping uses only half as much energy as crawling and is therefore better for moving from one place to another, such as returning to a safe refuge or going back to water.

Many species of frogs and toads are burrowers (Figure 21b), a mode of life that provides safety from predators, a source of underground invertebrates for food, and protection from excessive heat and dryness. Most of these use modified hind legs for burrowing backwards, something no other vertebrates do. The balloon frog, for example, is a bulbous shaped anuran that lives in the forests, savannah, and marshes of the lowlands of Burma and Thailand. It digs a permanent burrow in sandy banks and damp soil by alternate sideways movements of the hind legs. Each hind foot has a tough tubercle on the underside which helps it to move the soil sideways and compact it. The front legs are extended sideways to anchor the body against the side of the burrow as the frog moves deeper and deeper into the ground. Being a delicacy that is barbecued and eaten whole by the local human population, the main predator the balloon frog burrows to escape from is humans.

A few species of anurans burrow in a quite different way, going head first. The marbled snout burrower is widespread throughout southern and central Africa. Its other name is the shovel-nosed frog, on account of a pointed head which can pivot up and down on the vertebral column to help the digging. It usually starts digging using the hind legs, but then turns around to use its strong front legs for the main digging action, while the hind feet do the anchoring. Perhaps for them, coming across food in the burrow head first is important.

Well over 1,000 species of anurans are arboreal, living more or less permanently in trees and bushes. This is a relatively safe habitat because they can remain very inconspicuous, and the underside of the leaves provides shade and protection from drying out. Insects for food are likely to be abundant. Tree frogs belong to at least four different families, although there is a superficial similarity between most of them (Figure 21a). They tend to be small, well under 10 cm long, and the body rather flattened to help attach to leaves. The limbs are long, for reaching out to branches and twigs, and there are little sticky discs on the tips of the fingers and toes. Most of the tree frog species are well camouflaged, either green such as the European tree frog, or brown like the whistling tree frog of Australia. However, a number of them are amongst the most garishly coloured frogs of all, such as the Central American red-eyed tree frog, much valued in the pet trade. It has large, bright red eyes which it suddenly opens to distract potential predators, blue and yellow flashes, and vivid orange feet. The glass frogs are a group of South and Central American tree frogs living mainly in cloud forest. Their back and sides are green, but the under surface is sufficiently transparent to be able to see the body organs inside. No one has yet managed to explain why this should be so.

Being able to glide from a treetop to escape a predator, reach the ground, or move to another tree has advantages for an arboreal animal, and a handful of tree frogs have gone on to evolve the ability. Wallace's flying frog (Figure 21d) was discovered by the famous naturalist Alfred Russel Wallace in Malaysia. It is coloured bright green with a paler underside, and has the toe pads typical of tree frogs, but at 8–10 cm in length, it is large for a tree frog. Most of its life is spent in trees feeding on insects, and it only comes down to the ground to mate and lay its eggs. The front and back legs are exceptionally long and the feet are broad and webbed. These can be spread out sideways and, along with a flap of thin skin that can be stretched between the limbs, greatly increase the frog's surface area. On being disturbed, it jumps into the air and can glide or parachute to another tree up to 15 m away, or descend without harm all the way down to the ground.

Locomotion without legs: caecilians

Caecilians have no trace at all of limbs or limb girdles. They also have a very short tail, and in some it is completely absent so that the anus is right at the end of the body. Aquatic species such as *Typhlonectes* swim by lateral undulation of the whole body, helped by a slight side-to-side flattening and a fin running the length of the back. On land too, caecilians can move by lateral undulation like a snake. However, all caecilians are able to burrow, and for most of them this is their main mode of locomotion. The long, cylindrical body is marked by closely spaced circular grooves around the circumference, which increases the friction between the skin and the burrow. Many species have tiny, hard scales within the grooves, which gives the skin more protection against abrasion. Unlike the head of the other amphibians, that of a caecilian is pointed, compact, and very strongly built. Some of the bones of the skull are fused together, and there are no big spaces between any of them. Burrowing is by ramming the head through the mud or soil, using a curious mechanism called internal concertina locomotion. The skin is quite loosely attached. The body can be shortened within the skin by throwing the vertebral column into slight undulations, and then lengthened by straightening it. During burrowing, the skin provides anchorage against the sides of the burrow while extension of the vertebral column rams the head forwards. The head is moved up and down to push the loosened soil aside. Then the skin muscles move the skin forwards ready for the next shortening and lengthening of the vertebral column. In this way the animal digs a burrow that is scarcely any wider than its body.

Chapter 5
How amphibians feed

Adult amphibians are all carnivores, mostly eating invertebrates such as worms, slugs and snails, and insects. Larger species are capable of catching and swallowing small vertebrates, especially other amphibians and fish and also small lizards, birds, and mammals when the opportunity arises. Most of them have teeth which help keep the prey inside the mouth, but these are too small to be any use for chewing. Instead the prey is swallowed whole. The most effective way for an amphibian to feed when it is under water is to get as close to its prey as possible, then open its mouth suddenly and widely to suck the food in with a stream of water. This is easier than trying to just grab a swimming water beetle or minnow between the jaws. However, this kind of suction feeding is no use on land, because a stream of air is not dense enough to carry anything in. Amphibians feeding on land must use an alternative method. The most common is to have a sticky tongue that can be protruded from the mouth to capture the prey, then draw it into the mouth cavity. An alternative way to feed on land is to have strong jaws that can grasp the prey directly, like most reptiles. Amongst the amphibians, caecilians use this method.

Feeding in urodeles

When foraging in the water, a urodele such as a newt or salamander creates suction with the help of a special bone in the floor of the mouth called the hyoid. There are muscles in the throat than can pull it rapidly downwards to lower the floor, and this enlarges the mouth cavity. The jaws are opened at the same time, which increases the volume still more. It happens extremely quickly, and a current of water is drawn into the mouth, carrying the item of food with it. When the prey is small enough, such as a water insect, the suction alone is enough to take it in. But for capturing larger prey such as fish, or other amphibians, the urodele lunges its head forwards as its mouth cavity enlarges. Once inside the mouth, the food is held there by the fleshy tongue pressing it against the roof of the mouth and the teeth, while it is moved around and pushed backwards ready for swallowing whole.

The largest of all amphibians, the cryptobranchids such as the hellbender, feed only under water. They have particularly powerful oral suction and mostly take in larger food. During the summer this consists mainly of crayfish, but in the winter, when crayfish are less active, they rely on fish.

Larval urodeles, like frog tadpoles, use their respiratory current to also draw in food. The water is drawn into the mouth, again with the help of the hyoid bone, and the current flows backwards and exits through the gill slits. The flow is quite strong enough to suck in very small organisms such as water fleas that the larva lives on.

The 15 cm long yellow and black tiger salamander is the commonest of all the North American urodeles, ranging through woodlands and grasslands from coast to coast. Along with, for example, the similarly boldly patterned fire salamander of much of Europe, these are typical land-feeding salamanders, actively seeking out insects, earthworms, and small frogs. They rely on a muscular, protrusible tongue attached to the front of the mouth to catch their prey (Figure 22a). When one of them comes across an item of food, it presses its jaws against the ground and raises its head to open the mouth. The tongue is instantly extended out of the mouth by contraction of muscles attached to the hyoid bone,

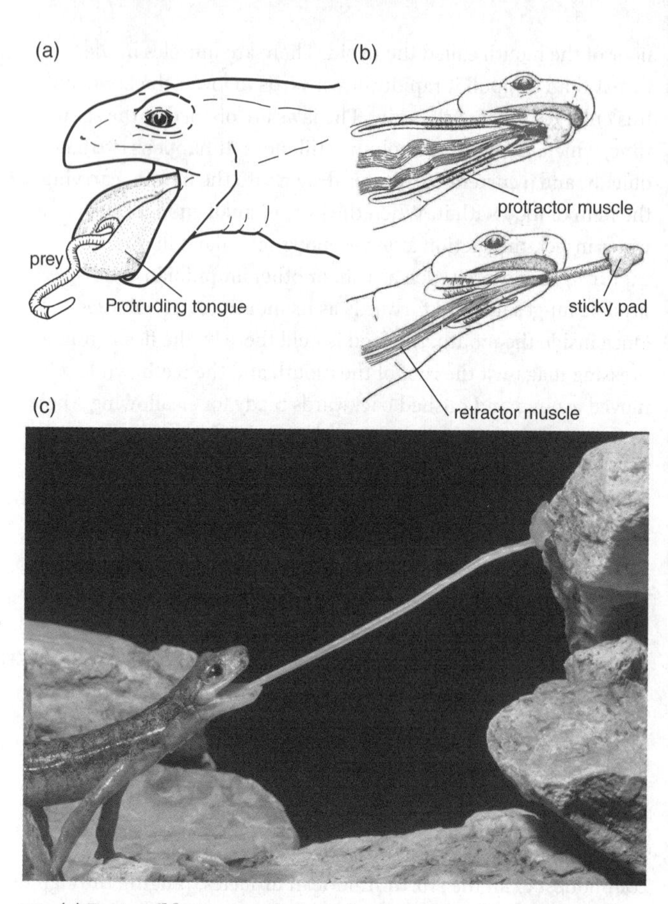

22. (a) Protrusible tongue of a typical salamander; (b) muscles of the extensible tongue of a plethodontid salamander; (c) plethodontid with tongue fully extended to catch insect.

and it becomes turgid with fluid forced into it. A sticky secretion covers the tongue's surface, the prey sticks to it, and the tongue is drawn back into the mouth, carrying the food with it. Small teeth along the jaws prevent it from escaping before it is swallowed whole. Despite the complicated muscles and bones involved, video footage shows that the whole sequence of prey capture and swallowing only takes about a fifth of a second to complete.

The plethodontids are the largest of all the salamander families, and one of their features is that they have lost their lungs. They manage without them because they are fairly small individuals that spend almost their whole life on land in a cool, moist environment. Under these conditions, their skin is sufficient to supply all the oxygen they need, without at the same time suffering loss of water. As a consequence of getting rid of the lungs, the mouth cavity was no longer necessary for breathing in air, and was free to evolve an amazing new way of catching food. The simple tongue of other salamanders developed into a highly effective projectile weapon, able to capture prey that is a long way away from the head (Figure 22b and c). The web-toed salamander of California, and the Italian cave salamander, can shoot the tongue out for as much as 80 per cent of the length of the body, although in most species the distance is more like half this.

The tongue is projected out by an ingenious mechanism. A special projector muscle winds around a set of hyoid rods, made of cartilage, that are embedded in the tongue. When the muscle contracts, it shoots the rods along with the tongue powerfully forwards out of the mouth. A second muscle, called the retractor muscle, connects the very front of the tongue all the way back to the pelvis. When the tongue lies inside the mouth, the retractor muscle is folded up, but projecting the tongue stretches it out. By then contracting it pulls the tongue, hopefully carrying the prey, all the way back into the mouth. A sticky pad at the tip of the tongue actually catches the prey. Not only is prey capture extremely fast, about 50 milliseconds, but extremely accurate. The two eyes are used together and the animal uses stereoscopic vision to judge the precise distance to its prey. Even a flying insect momentarily settled on a surface is given little chance of escape.

Feeding in anurans

Amongst the frogs and toads, only one group, the highly aquatic pipids such as the Suriname toad and African clawed frog, use simple suction for feeding. The floor of the mouth cavity is lowered, the food sucked in, and the mouth closed to expel the water while the prey is held by the teeth. For larger prey, and this often includes scavenged dead creatures as well as fish and other amphibians, the front legs have to be used to help push food into the mouth.

All the other anurans use a tongue for catching their prey, although the mechanism is very different compared to salamanders, and evolved separately. The tongue is attached at the front of the mouth, and when it is not being used it is folded backwards over the floor of the mouth cavity (Figure 23a). When food is encountered, a muscle inside the tongue contracts making it stiff. Another muscle, connecting the tongue to the front of the jaws, then contracts and pivots the tongue forwards out of the mouth, and downwards onto the prey. The tongue has a pad on its end covered in mucous to which the prey sticks (Figure 23b). Finally, a retractor muscle pulls the tongue back into the mouth along with the prey, which is held by the small teeth to stop it escaping. Swallowing is helped by the large, bulbous eyes pushing downwards on the buccal cavity.

There is not very much variation in the tongue feeding mechanism amongst different anurans. Some can extend the tongue further than others, and some, such as cane toads, have much stronger jaws for feeding on larger prey. The most specialized tongue is found in burrowing species, for it is not easy to use tongue flipping in a tight burrow. The Mexican burrowing toad feeds on the ants and termites that it comes across underground. Its tongue is flat when it is inside the mouth, but can be rolled up like a tube. Instead of being flipped out, it is extended by the pressure of the

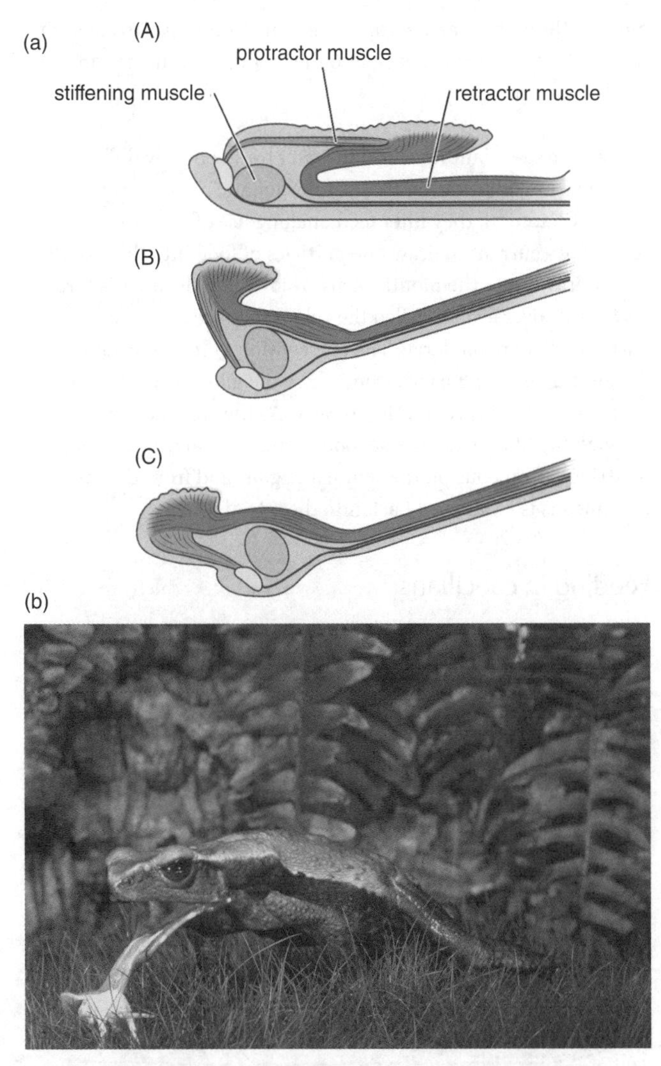

23. (a) The mechanism of tongue protrusion in an anuran. The stiffening muscle acts as a fulcrum, the protractor muscle flips it forwards out of the mouth, and the retractor muscle pulls it back in; (b) Protruded tongue showing sticky terminal pad.

muscles through a narrow gap at the front of the jaws to catch the insects. It is then drawn back into the mouth, carrying the insect stuck to its tip.

Tadpole larvae of anurans feed in water by an entirely different method from adults, even those of the aquatic pipids. As we have seen in Chapter 3, they filter feed, making use of the existing respiratory current to draw fine particles of food into the mouth. Water drawn into the mouth cavity by lowering the floor is forced backwards over a valve called the velum that keeps out any particles that are too large. The current flows into the front part of the gut called the pharynx, from where the gills open into a gill chamber and out through the spiracle. As the current flows through the pharynx, any fine food particles it carries are trapped by strings of mucous on the velum and gills, and from time to time the mucous is swallowed back into the intestine.

Feeding in caecilians

The caecilians are adapted for feeding on land but in a completely different way from other amphibians, one much more like the way reptiles take in food (Figure 24). Instead of a protrusible, sticky tongue, they grasp their prey directly with their strong jaws and teeth. The head is solidly and strongly built for burrowing into the

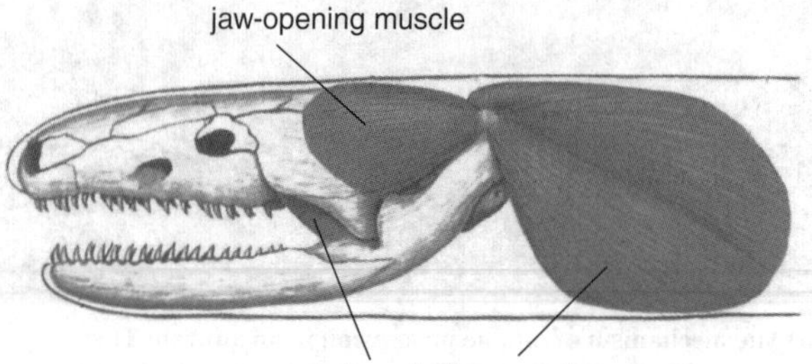

24. Caecilian skull, teeth, and jaw muscles.

ground or the bottom of the water. This reduces the amount of space there is for normal jaw-closing muscles on the side of the head. But the force of the bite has been hugely increased by the evolution of a large extension of the lower jaw behind and above where it hinges on the skull. It is called the retroarticular process, something none of the other amphibians has. A massive muscle attaches to it and runs backwards alongside the neck. The jaws themselves are lined with sharp teeth that pick up the food, and by sloping backwards they help to prevent it from escaping from the mouth.

The caecilian diet consists mainly of earthworms, termites, and other soft invertebrates that it comes across in its burrow. One of the most curious feeding activities seen in at least some species of caecilian is the habit of rotating the whole body in the burrow, while hanging on to a newly captured worm with its teeth. This is an effective way of tearing up an item of food such as a large earthworm, that is too big to ingest whole.

Chapter 6
The amphibians' world: sense organs and communication

An animal's sense organs send signals to the brain, which interprets their messages to create a representation, or picture, of the current world around it. This includes detection of food, and the presence of hazards such as physiologically unsuitable conditions or approaching predators. It also includes the presence, activities, and social interactions of other members of its own species, like rivals for territory and potential mates on the part of males, and signals showing readiness to mate on the part of females. The way that parents of many amphibian species protect their young depends on detecting signals given out by the offspring.

Amphibians have the eyes, ears, olfactory organs of smell in the nose, and touch receptors common to all vertebrates, but the relative importance of the different senses varies from group to group. To some degree this is related to the particular habitats and modes of life of different species. As a very broad generalization, the anurans have a sensory world most like our own. Their vision is good, and includes the ability to see colours, and their hearing is acute. In contrast to this, the urodeles rely much more on their senses of smell and touch. This is even more true of the caecilians, which spend most of their time living in a burrow with no light at all. Amphibian larvae have an additional sensory system called the lateral line system. A set of fine grooves run over the head and body, and house little clusters of sensory cells called neuromast organs. They respond to vibrations in water caused by the movement of nearby swimming animals. The lateral lines are lost during metamorphosis, apart from in the highly aquatic species like pipid anurans and the paedomorphic urodeles, where they are still important for detecting prey and predators in the surrounding water.

Amphibians use several sensory cues in combination to navigate around their territories. They can recognize visual landmarks they have learned such as the shapes of trees and rocks, and characteristic odours and sounds. Many species of both urodeles and anurans migrate annually to and from the same breeding site. This is sometimes a matter of travelling several kilometres at night, which demands an accurate geographical homing ability. Some, for example the black and green marbled newt of north-eastern Europe, can navigate home by the stars. Like sailors, to achieve this they must have the equivalent of a star map and a clock in the brain, although we do not yet know very much about these. Careful experiments show that other species, such as the alpine newt and the North American red-spotted newt, are sensitive to the Earth's magnetic field, which they can use to determine the direction they must take.

Anurans

The eyes are the most important sense organs of anurans for finding and catching their food, and avoiding predators. The habitat of a typical frog demands particularly versatile vision, because the animal spends part of its time viewing its surroundings in air, and part while swimming under water. Furthermore, it has to cope with a wide range of light intensities, from bright daylight to night time. Frogs' eyes have the familiar bulbous shape, positioned on the top of the head rather than the sides, as loved by cartoonists. This gives the eyes more or less complete all-round vision, and when swimming in the water, or just resting at the surface, a frog can view its complete surroundings and detect anything moving within it. But the two eyes can also be brought forwards to look at the same object. This binocular vision, as it is called, allows the distance to the object to be accurately judged. The transparent lens inside the eye that focuses the light on the retina is nearly spherical. Therefore, unlike our own lens, it is strong enough to focus the image of an object even while the frog is under the water. But the lens can also focus accurately in air, using special muscles that pull it backwards and forwards. Also unlike our own eyes, there is no fovea or yellow spot, a highly sensitive part of the retina on which the image is focused to see fine detail. Instead, the eye is extremely sensitive to any movement in its visual field. When hunting, for example, a frog interprets as food anything moving that is small enough, and will chase and catch it by continuously tracking its movement with its eyes.

Anurans have excellent colour vision, even better in some ways than our own, and hardly surprising in a group of animals where colour plays such a big role in their behaviour. Vertebrates have two types of light-detecting cells in the retina. The rods are most sensitive to movement, and they operate in low light levels. The cones are responsible for colour vision, and are only sensitive at higher light intensity. Different cones are stimulated by different wavelengths, or colours of light, and the anuran retina is described as trichromatic because it contains three types, sensitive to green, blue, and red light respectively. The eyes therefore have a wide range of colour discrimination. Alone amongst vertebrates, frogs also have two different kinds of rods in the retina, one sensitive to blue light and the other to green. What this means is that their eyes can see certain colours at much lower light levels, where other vertebrates only see in black and white. This ability serves the frog well when it is active at night. There is yet another unusual kind of rod in the retinas of tadpoles, and also of highly aquatic species such as the pipids. They are sensitive to ultraviolet light, which travels further than lower frequency light through water, thanks to its higher energy.

The important use of coloration by amphibians for camouflage, and for warning off potential predators was discussed in Chapter 1. Many species of frogs also use colour for communicating with members of their own species. One common way is to have a brightly coloured vocal sac, which creates a conspicuous orange, yellow, or white signal when inflated. Sometimes this is used as an aggressive signal towards other males. In other cases, it is a signal to attract a female. Males of the European common frog, for example, have a bright white sac that distinguishes them from the females during the breeding season. A similar use of colour during mating is found in the European moor frog. This is a species with what is called a 'scramble' mating system, in which several males simultaneously try to mate with the females. At the start of the mating season, the males all turn a bright blue colour, while the females remain dullish brown. The distinction prevents males from accidently trying to mate with one another.

A few anurans indulge in a different kind of visual communication, called leg-waving. The male Indian dancing frog (Figure 25) has a bright white vocal sac for displaying towards a female, but he also attracts her attention by tapping his front legs and then sticking out and waving his conspicuously coloured, webbed back legs. These, and other leg-waving frogs, live close to fast-running water, which can drown out the sound of vocal communication. Leg-kicking movements are also used sometimes as aggressive signals towards other males, and may even be used to physically kick a rival away.

Hearing is also an important sense in anurans, and of the three amphibian groups, they are the only ones that have an ear drum (Figure 26). This is a large, thin sheet of tissue visible on the side

25. Indian dancing frog leg-waving.

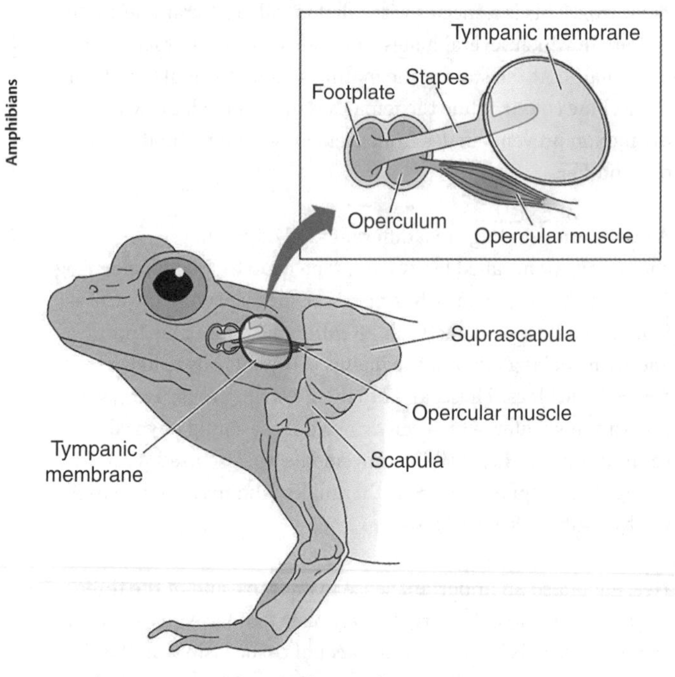

26. The structure of the middle ear of an anuran.

of the head behind the eye. When sound waves travelling through the air reach it, they force it to vibrate at the same frequency as the sound. The vibrations are passed on to a long, slender bone attached to the ear drum, called the stapes. The other end of the stapes fits into a small hole in the side of the skull. This, the oval window, is the outer opening of a chamber called the labyrinth that lies inside the braincase and is filled with a watery fluid. Finally the labyrinth has a patch of sound sensitive hair cells embedded in a part of its wall known as the basal papilla. When the stapes passes the sound vibrations from the ear drum into the fluid inside the labyrinth, they cause tiny movements of these delicate hair cells. This in turn stimulates the generation of nerve impulses in the sensory nerves. The nerves pass the messages on to the brain, where they are interpreted as sound. The important point about the ear drum of the anurans is that sound is collected over a large area, and then concentrated by the stapes at the far smaller oval window. This increases the pressure of the sound waves entering the labyrinth, making the hearing much more sensitive, especially to higher frequencies. Experiments in the laboratory have shown that a typical anuran can hear sound up to around 1400 Hz. In our terms, this is a frequency about two octaves above the middle C of a piano. This is rather poor by the standards of most mammals and birds, including ourselves, but is certainly good enough for the relatively simple vocal communication of frogs.

Anurans have a second way of hearing, which does not depend on an ear drum, but detects lower frequency sound (Figure 26). This, importantly, includes sound carried through the ground, such as the noise of the footfalls of an approaching predator. There is a little bone called the operculum, lying next to the oval window and attached by a muscle to the shoulder girdle. Sound waves passing through the ground are picked up by the front legs of the frog. They are transmitted via the shoulder girdle, the muscle, and the operculum into the labyrinth. Here a second sensory patch of hair cells in the labyrinth wall is stimulated. It is called the amphibian papilla, because it occurs in other amphibians but not in any other tetrapods.

We met the huge role played by vocalization in the courtship of anurans in Chapter 3. The noise they make may be very loud, and when calling in chorus can often be heard a kilometre away. The loudest species is said to be the common coqui frog of Costa Rica. Its name comes from the high pitched 'co-kee' sound of the call, which it makes all night, from dusk to dawn, during the breeding season.

Frogs create their call by filling the lungs with air and then contracting the muscles of the abdomen. This forcibly expels the air across the larynx, which contains stretched vocal cords that vibrate. The frequency of the sound made by an individual can be varied by altering the tension of the vocal cords and the rate of air flowing over them. This increases the repertoire of different calls it can make for different purposes. In the great majority of anuran species, the sound produced by the vocal cords is amplified and radiated outwards by a vocal sac, or pair of vocal sacs (Figure 27). They consist of the skin of the throat, plus a thin layer of muscle tissue, and are connected to the mouth cavity by an aperture in the floor of the throat. The pressure of the air entering a vocal sac inflates it, often to an impressive size, as a series of calls are being emitted. After calling, the sac is deflated by the elasticity and muscular contraction of its walls.

It would be almost impossible to use vocal sacs underwater, and the aquatic pipids such as *Xenopus* use an entirely different method of calling during courtship. They have neither vocal cords nor vocal sacs, but make a series of short clicks by opening and closing a pair of special cartilages in the larynx.

A measure of the importance of vocal communication in the life of frogs is that it is often the most energetic thing they do. A continuously calling individual uses about 50 per cent more

27. Marsh frog with a pair of expanded vocal sacs.

oxygen than it does for even the most energetic locomotion, and its metabolic rate can rise to over ten times the resting rate. Yet many species continue for hours on end at night. The payoff for using all this energy is that the more vigorously an individual calls, the more likely it is that he will attract a mate.

There are several reasons for communicating vocally, which has the two advantages of being effective over a distance, and in the dark. The commonest calls we hear are known as advertisement calls. One of these is to indicate to other males that the caller already occupies this territory. Another, the aggressive call, is made by a male when it finds itself close to a rival, and wants to warn him off, hopefully without combat. Advertisement calls are also made to attract the attention of females, and try and persuade them of his suitability as a mate. When close to a female, a male may make another kind of call to help her actually locate him. In the presence of danger, such as an approaching predator, many species give out a very loud warning or distress call to alert other individuals in the vicinity. The Australian green tree frog, for example, is said to sound like a distressed cat when threatened, and its call may be so loud that it startles a predator that has just caught an individual into dropping it.

Urodeles

Most urodeles are silent, and sound plays no part in their social communication. One of the few exceptions is the Pacific giant salamander, which gives out a sound like a dog barking as a warning call when disturbed. Also urodeles have no ear drum, and therefore cannot even hear the higher frequencies of sound that anurans can. However, they do have the second frog-like hearing system, with an operculum connected by a muscle to the shoulder girdle, and an amphibian papilla. With this they can detect sounds of lower frequencies, both airborne and ground-borne via the front legs. This warns them of a number of things likely to be significant, such as the approach of a large predator or the sound of splashing water.

Vision is also less dominant in the life of salamanders than in anurans. Their eyes are adapted for mostly low light conditions, and they cannot focus very accurately. Despite this, sight is still the main sense most urodeles use for detecting and capturing their prey. As we might expect, the lungless plethodontid salamanders with their long, extensible tongue have the largest eyes and the most acute vision. They rely on binocular vision, using the two eyes together, to estimate the distance to the prey. In common with most vertebrates living in dark caves, the eyes of several species of permanently cave-dwelling urodeles have been reduced in size and function. The Texas blind salamander's eyes are no more than small black dots. Those of the olm are tiny and covered over by skin, and unable to form an image at all, although they are still sensitive to the general light level.

The role of vision in social behaviour is based on whole body postures and movements. A male red-backed salamander expresses increasing levels of aggression towards another male intruding on his territory by raising his body off the ground and increasingly arching his back upwards (Figure 28). Submission by the intruder is indicated by pressing its body against the ground and turning its head away. A male trying to attract a female will wag his tail, or swing his front legs, called the butterfly display. The crest of the European great-crested newt is a familiar visual signal. It runs all the way along the back and tail, and grows more prominent in the spring when the newts emerge from their winter refuge and enter the water. This species is a lek-breeder, which means that numerous males occupy the same area, such as a shallow part of a pond. Here each one defends his own small territory while attempting to attract a female by signalling with his tail and flashing the white stripe along the body.

The sense of smell plays a very important role in urodele behaviour. Given their relatively poor eyesight, it is particularly necessary for detecting stationary prey, and for recognizing the home site. Smell also plays a leading role in territorial, aggressive, and courtship behaviour. Like most tetrapods, urodeles have two chemical-sensing organs. The main olfactory organ lies inside the nose, and is used to detect airborne molecules carried in the stream of air breathed in. The second is called the vomeronasal organ, or sometimes Jacobson's organ after the person who first discovered it. This one lies just inside the nostril, and is mostly used to smell molecules that are picked up directly from the ground by the upper lip. In the plethodontid salamanders, there is even a special grove leading from the lip into the nostril, along which the molecules pass.

Molecules produced by an animal specifically for communicating with one another are called pheromones. They are especially important in territorial behaviour and courtship of salamanders.

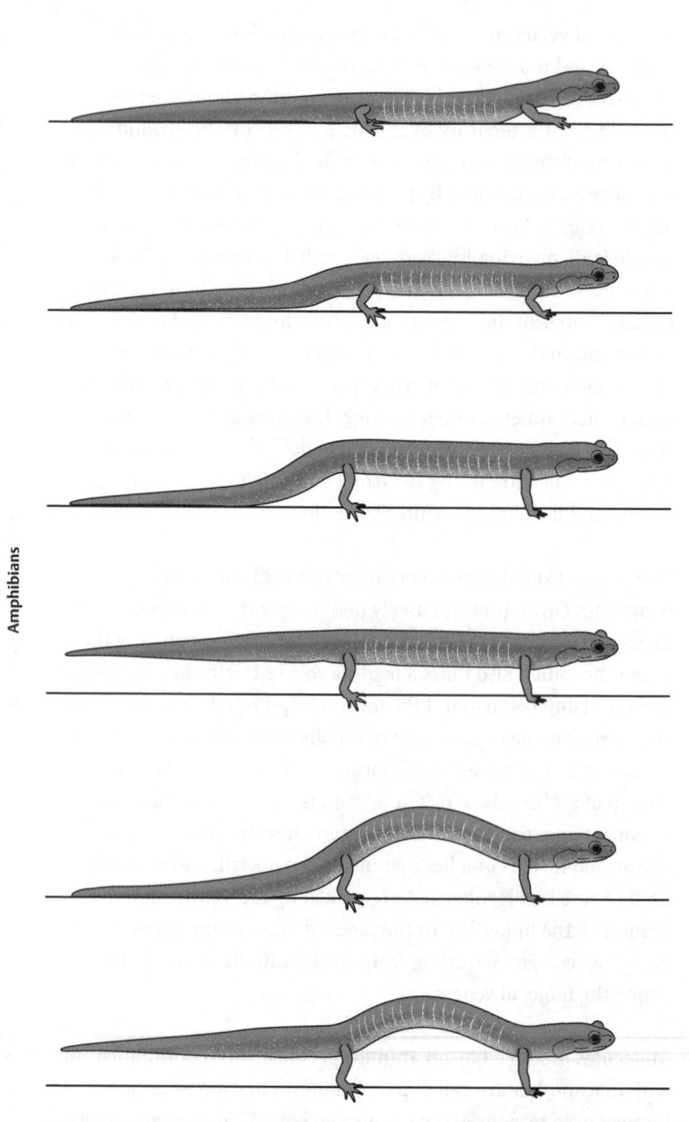

28. Sequence of increasingly aggressive visual poses of the red-backed salamander.

Glands around the cloacal opening produce certain pheromones that are spread on the faeces left behind to mark the animal's territory. Any intruder into the territory is recognized by its pheromones, and is likely to be attacked and then chased off by the owner. Pheromones are also indispensable in courtship. In the much-studied red-legged salamander of North America, for example, a male recognizes a female by her smell, approaches her, and touches her with his snout. He secretes a pheromone of his own from glands around his chin, which stimulates the female. He may even use his teeth to puncture her neck and introduce pheromones directly into her body. Thus stimulated, the female starts courtship by straddling the male's tail, which in turn stimulates him to deposit a spermatophore.

Caecilians

Not very much is known about the sensory world of caecilians beyond what can be told from the nature of the sense organs. Vision is clearly unimportant because the eyes of all species are very small, and as we know from those few studied in the laboratory, they are no use for activities like foraging or mate recognition. At best the eyes can distinguish light from dark as an aid to burrowing, while a number of species are totally blind and the eye covered by a layer of skin or bone. Similarly with hearing. Like urodeles, caecilians do not have an ear drum and probably the main ability of the ear is to detect ground-borne sound waves that reach them in the burrow.

The most important caecilian sense organs are a pair of short tentacles on the head, which open through a small hole between the nostril and the eye (Figure 29). A muscular sheath surrounds them so that they can be extended or retracted. They connect to the vomeronasal organs within the nostrils, and are sensitive chemoreceptors, for example detecting molecules associated with food in the burrow. Surprisingly, the tentacles seem to have evolved from eye tissue no longer needed for eyes, because they

29. Caecilian head showing the sensory tentacle and reduced eye.

can be moved by what is normally an eye-moving muscle. In one species, there is even a rudimentary eye at the tip of the tentacle, which may have a simple visual function. Given the very limited hearing and virtual absence of vision in caecilians, it is more or less certain than pheromones play the main role in social behaviour, although practically nothing is yet known about this.

Larvae and the lateral line system

All fishes have a sensory system called lateral lines, and so do larval amphibians. They are grooves in the skin of the head and body that are sensitive to the movements of nearby organisms in the water. Small groups of cells called neuromast organs are embedded periodically along the groove, and each one connects to a little projection called a cupula. A small pressure difference between different points along the lateral line creates a minute movement of water. This moves the cupula, and the cupula in turn stimulates a sensory nerve cell. The pressure difference can be caused by a nearby swimming animal, or by a reflection of the larva's own movement off a solid object or the river bank. Lateral lines are mainly important for detecting and evading predators.

Most species lose their entire set of lateral lines when they metamorphose into adults, but those amphibians which continue to be mainly aquatic keep them throughout life. For example, adult pipid anurans, and the paedomorphic urodeles like cryptobranchids and sirens, find them useful for detecting potential prey swimming nearby.

Chapter 7
Amphibians and humans

Amphibians in myth and folklore

With a little imagination the squat body, long legs, bulbous head, and prominent, staring eyes make a frog or toad look like a strange human. With such a form, plus their secretive habits, it is no wonder that frogs and toads have figured in imaginative human minds trying to make sense of the world, sometimes as good, sometimes as evil players. In ancient Egypt, the annual flooding of the River Nile, essential to fertilize the fields for cultivating the following year's crops, was always accompanied by the appearance of huge numbers of frogs. People worshipped the water goddess Hekel, who was endowed with the head of a frog. The frog was also the symbol of the Egyptian goddess of human fertility, called Heqit. Women would often wear an amulet in the shape of a frog in order to win her favour and become pregnant.

In ancient China, the toad was a rather more ambivalent character. One legend had it that toads possessed the secret of eternal life. A wise man called Liu Hai wandered the world with a toad called Ch'an Chu as his companion, who told him the secret. To the early Mesoamerican Olmec civilization, the toad was the god of rebirth, a role suggested perhaps by the way it would shed and eat its own skin, as if symbolically consuming itself and re-emerging.

In European folklore, toads were the companions of witches, and often up to no good on this account. People were wary of meeting a toad, in case it was really someone in disguise who it would not be a good idea to offend. There were some very elaborate rituals for using toads to destroy a witch's power. One example from Devon, England, was for the person affected to place the heart of a toad pierced with thorns, along with the liver of a frog, in each of three jars. These were sealed and buried seven feet down and seven feet from the church door, in three different churchyards, at the same time as repeating the Lord's prayer backwards. Thereafter, the individual was free of the influence of witchcraft.

No one knows where the curious idea of toad jewels, or toadstones, comes from. This was the widespread belief, referred to for example by Shakespeare and John Bunyan, that a toad has a jewel in its head which, when carried around, acts as an antidote to poison. Certainly there were plenty of rogues at country fairs happy to pass off more ordinary stones as these magical items. The idea probably arose because a toad is, of course, immune to the poisonous nature of its own skin. Indeed, the toadstone was just one of many beliefs in the curative properties of anurans. Hanging frogs and toads around the neck was believed to cure many ailments, while eating a live frog, or a stew made of frogs was used as a cure for whooping cough. As we shall see, the idea that anuran skin products have medicinal properties is not unfounded.

The most enduring and endearing stories of frogs and toads are those fairy tales in which for one reason or another a beautiful, kind princess has to kiss or marry the animal, whereupon it promptly turns into a handsome prince. The story occurs in one form or another throughout European culture. It offers the morally pleasing conclusion that kindness and understanding by beauty towards the unfortunate reaps the reward of happiness ever after.

Urodeles also figure in a number of myths. Salamanders were believed to be immune to fire, probably because they frequently rest and hibernate in rotten logs. When such logs were brought into the house and thrown onto the fire, the animal emerged apparently unharmed. The Jewish Talmud mentions that smearing the body with the blood of a salamander protects a person from fire, and in China, robes allegedly made of salamander skin protected the wearer from fire. Most likely they were actually woven from asbestos fibres. Another belief, widely held in France, was that salamanders were extremely poisonous, and would poison the water in a well, or even poison the fruit of an apple tree simply by climbing up it.

One of the oddest stories about urodeles concerns a 1 m long fossil, now called by its scientific name *Andreas scheuchzeri* (Figure 30). When first discovered by the Swiss naturalist Johann Scheuchzer in 1726, he believed it was the remains of an unfortunate human being that had drowned in the biblical flood, and accordingly he named it *Homo diluvii testis*—'the man who witnessed the flood'. We now know that it is really the fossil skeleton of a giant cryptobranchid salamander.

Amphibians as food

As well as creating myths and folklore about amphibians, we humans have always made use of them, as we have of so many other organisms with which we share our world.

One estimate is that worldwide between three and four billion frogs a year are eaten, usually the legs for no other reason than that they have the most muscle tissue. As well as local sources, there is a large export market, especially from China and Indonesia, thought to be between half a billion and a billion individuals per year. Frog meat has a delicate, chicken-like taste,

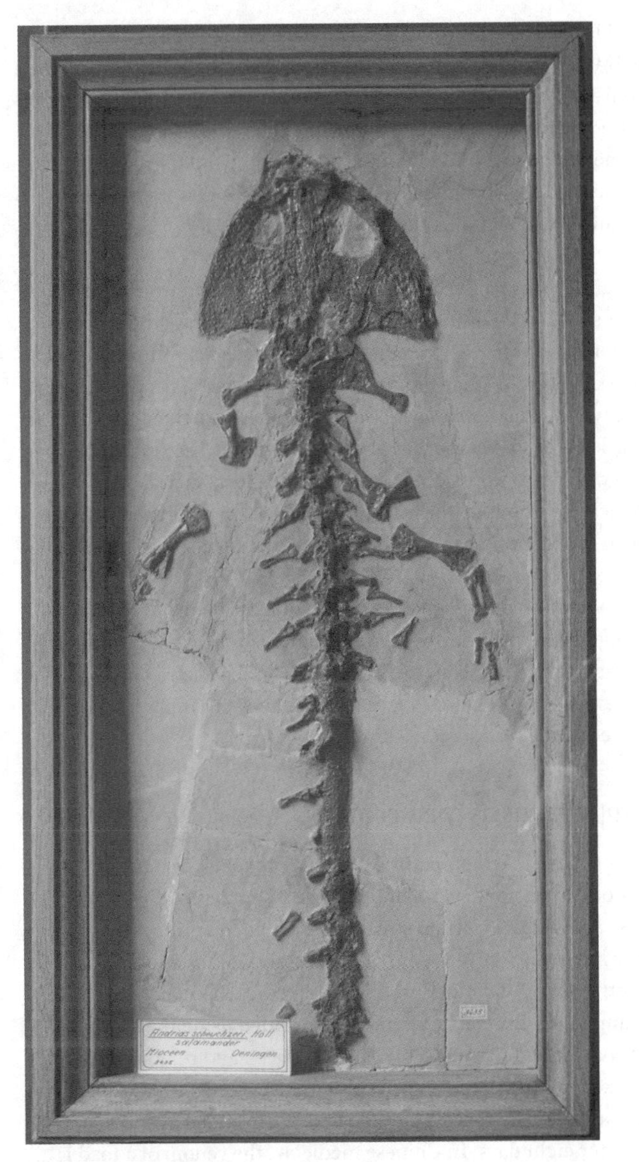

30. ***Homo diluvii testis.***

and is particularly rich in the beneficial omega-3 fatty acids and vitamin A. In Europe, especially France, frogs' legs are a well-known delicacy. In China frogs are called field chickens, and a traditional braised frog dish made of the whole animal carcase is sometimes served. Indonesians make a soup out of frogs' legs and fermented soya beans called swikee, which is definitely an acquired taste for outsiders.

Originally, it was European edible frogs collected from the wild that satisfied the European demand, but nowadays the larger American bullfrog is bred on frog farms specifically for the food market. In Asia, huge numbers of these are now farmed, although many individuals of wild species are still collected for sale at local markets. In some other parts of the world too, frogs taken from the wild make up a significant part of the local diet. In West Africa, for example, the huge, 3 kg goliath frog has been eaten to such an extent that the species is now classified as endangered.

The water-holding frog of Australia is traditionally used by indigenous Australian people as a source of drinking water in arid desert conditions. They are dug up from their metre deep burrows in sandy soil and gently squeezed until they give up the water stored in the bladder and skin cavities.

Amphibians as medicine

People have relied on natural medical products for most of their history. Even now, over half the world's population does not have easy access to modern medical facilities, and continues to rely on folk medicine for treatment of disease and injury. Amphibians contribute a good number of such remedies. In Spain, for example, traditional folk remedies included using the ashes of a cremated toad to treat arthritis and cancer, and remove warts. Snake bites were treated by rubbing the wound with a whole toad, and anthrax supposedly cured by tying a live toad over the entry sore for eight days. In Chinese medicine, the venom of a toad is the basis of a centuries old anti-cancer remedy called chan su. It comes in the form of small brown cakes to be eaten, and its use is regulated by the Chinese pharmacology authorities. The list of ailments treated in Brazil by toad skin, venom, or the whole body is extraordinarily long. It includes arthritis, asthma, backache, cancer, cough, diarrhoea, earache, infections, inflammation, influenza, osteoporosis, rheumatism, sore throat, strain, and toothache.

These traditional beliefs in the effectiveness of amphibians for the treatment of various diseases include some ideas that are quite absurd to modern scientific thinking. No doubt any actual beneficial effects were due to faith in the treatment itself rather than to any actual pharmacological action of the substance. This is what we nowadays call the placebo effect, and of course it can be remarkably effective.

However, the skin and the salivary glands of amphibians, especially anurans, do produce a large range of molecules that have real biological effects, and modern research is beginning to show that by no means all the folk remedies are unfounded. There are several very promising discoveries of substances that have potentially important medical properties. The toxins of the yellow-bellied toad, and also *Xenopus*, include molecules with antibiotic properties that act against a number of pathogenic bacteria, including the human gut pathogen *Escherichia coli*. They are similar in effectiveness to antibiotics such as penicillin, and offer a possible source of new antibiotic drugs in place of those where drug resistance has become such a problem due to overuse. Equally exciting, the glandular products of several species of toads have been found by rigorous laboratory tests to inhibit the growth of colon cancer, melanoma, leukaemia, and other cancer cells. Another potentially important case is a toxin of the South American poison dart frogs called epibaditine. It is a very powerful painkiller, even more effective than morphine, but apparently without the problem of addiction that morphine causes. The paradoxical frog, long noted for its giant tadpoles as described in Chapter 3, gained more recent fame for an antibacterial skin product found to stimulate insulin production in humans. A synthetic version is already used to treat type 2 diabetes.

At the moment, few of these substances have been tested enough to see whether they have unacceptable side effects, but even if they have, they may be modifiable in the laboratory to become useful drugs. It is often argued that one of the good reasons for conserving plant and animal species is that they may prove to have as yet undiscovered medical benefits. Of no other group of animals is this more true than amphibians.

Amphibians as pets

Quite why the keeping of pets, such as cage birds, small mammals, mantids, and crickets, is so universal in human societies, especially urban ones, is not at all clear. Perhaps it is simply a deep-seated affinity with the natural world amongst city people who have lost intimate contact with nature. Whatever the reason, amphibians are much sought after and harvested from the wild for this purpose, in sometimes prodigious numbers. The little bright yellow, orange, or red golden mantella frog is a case in point. These only live in a few areas of mountainous swamps in east-central Madagascar, from where they are heavily collected for export. Between 1994 and 2003, at least 230,000 mantellid specimens, half of them the golden mantilla, were exported. Hong Kong is amongst the largest markets for the amphibian pet trade, both for importing and exporting the animals. In one year, it received a recorded 34,000 oriental fire-bellied toads from South Korea and 9,350 Japanese fire-bellied newts, amongst many other species. The exports mostly go to the American and European markets, where they are sold as pets at a very large profit, in a significant number of cases illegally.

Amphibians regularly harvested from the wild for the pet trade, both legally and illegally, are mostly rare, brightly coloured species, such as the mantellids just mentioned, the South American poison-dart frogs, and the Central American red-eyed tree frog. Other unusual species are in demand, such as the world's largest frog, the goliath frog. Several hundred of these are exported each year from the coastal rain forests of West Africa to the USA, where they fetch about $3,000 each. The Chinese giant salamander is similarly much sought after.

Amphibians and poison darts

Coating the tips of darts or arrows with a highly toxic poison to make them lethal is common worldwide amongst indigenous people. The main poisons used come from plants, such as curare in South America and oleander in Africa. But several tribes of western South America use toxin from the skin of the poison dart frogs, a few vividly coloured members of the genus *Phyllobates*. The toxin is collected by heating the frog over a fire, or piercing its body with a stick, which makes the skin exude its glandular contents as a froth. The darts are prepared by dipping their tips into the froth. The exception is the golden dart frog of western Columbia. It has the most highly poisonous toxin of all, so much so that the dart only needs to be rubbed directly on its skin without harming the frog. Once prepared, the darts are blown out of a hollow plant stem into the flank of the prey, where the toxin rapidly begins to work. The animal is followed until it finally collapses in front of the hunters.

Surprisingly, the frogs do not manufacture the toxin themselves, but acquire it from certain ants and millipedes that they eat. The tadpoles of poison dart frogs are also highly toxic. In their case, they are provided with the poison by the mother, through the unfertilized eggs she lays for them to eat. The poisons themselves are called batrachotoxins, members of a group of chemicals called alkaloids that includes the deadliest biological toxins of all.

A single dart rubbed on the back of a golden dart frog has enough to kill a monkey in seconds, and even a large mammal such as a horse. The total amount of the toxin in the skin of one 3.5 cm long frog is enough to kill as many as 22,000 mice, or 10 humans. Batrachotoxins act in a number of ways on the tissues of the body, mainly by preventing nerve conduction and causing heart failure. The only animal apparently immune to the poisonous skin, and therefore their only natural enemy, is the frog-eating snake.

Exploitation of amphibians gone wrong: the cane toad

With a body 10–15 cm long, the cane toad (Figure 31) is a very large toad that was originally restricted to the Americas, from Texas in the north to Peru and the Amazon Basin in the south. It is a voracious feeder on small vertebrates as well as invertebrates, and will even eat carrion. Beetles are a major pest of sugar cane crops, and early in the 20th century cane toads were introduced into Puerto Rica in the hope of eradicating a serious infestation.

31. Cane toad consuming another frog.

This was successful, and as a consequence cane toads were introduced into several Pacific countries such as Fiji and the Philippines, and Caribbean Islands. They were also introduced into the sugar cane fields of Queensland, Australia, in the 1930s. Unfortunately, the particular beetle species attacking the canes in Australia spent most of its time at the top of the cane plants, where the cane toads could not get at them. Nevertheless, the cane toad population rapidly grew, and has become a serious threat in its own right to the wildlife of the eastern regions of Australia. This is in part due to their very broad appetite for small native mammals, reptiles, and other amphibians, and in part to their very high rate of reproduction: a female can produce 20,000 eggs. But the main problem with cane toads, and also with their tadpoles, is the highly toxic skin. They are lethal to most of the Australian carnivores that try to eat them, such as the quolls, monitor lizards, snakes, and crocodile. Domestic dogs that attack them die too. In fact their only serious predators are some of the opossums, which seem to be able to devour them safely.

Amphibians in teaching and research

Nothing evokes an image of traditional school biology teaching more than dissecting a frog. Here is an animal that can be easily killed and pinned on its back to a board, and the thin, loosely attached skin of the belly cut through by a pair of scissors. With neither ribs, thick abdominal muscles, or excessive fat to get in the way, the colourful internal parts are soon exposed—dark red liver, green gall bladder, bright red blood vessels, pink kidneys. Even better, by covering the dissection with water the organs all gently float apart from each other, so there is no clearer way of revealing the internal anatomy of a vertebrate. Their widespread use was also a measure of how readily and cheaply frogs were available in large numbers. Literally millions of leopard frogs caught in the wild used to be sold each year by suppliers in the USA, and equally prodigious numbers of common frogs in Europe. The huge fall in the numbers of these species, caused in no small part by such harvesting, has led to much greater sensitivity to conservation. Most specimens are now bred for the purpose, and they are not used so wastefully. Nowadays, a class or group demonstration dissection is likely to replace the provision of a specimen for every student to conduct their own.

Frogs and other amphibians have also proved to be extraordinarily valuable in a number of areas of biological research. They are easy to dispatch humanely by destroying the brain with a needle inserted at the back of the head, a method called pithing. This leaves the other tissues and organs—skin, muscles, kidneys, intestines, etc.—undamaged and much easier to conduct experiments on than those of other vertebrates such as mammals. For example, amphibians can live for long periods submerged in freshwater, but to prevent too much loss of salt from the body, the skin must be able to actively take it up. The epithelial cells pass sodium and chloride ions from the very low concentration in the surrounding water into the blood, which has a much higher concentration. The mechanism is called a sodium pump, and it takes a lot of energy to push the ions against the concentration gradient. The first understanding of how chemical energy is harnessed to drive the pump came from the study of frog skin. This was of fundamental importance because ion pumps occur in many other parts of every animal, including of course humans. In another area of research, work on the frog eye using visual stimuli and electrical recording of retinal cells first showed how visual information is processed by the retina before being passed via the optic nerve to the brain. The frog kidney is not the compact structure found in other tetrapods, but a much flatter, more diffuse organ, and the contents of individual kidney tubules that produce the urine can be much more easily sampled using a micropipette. Almost 100 years ago, the concentration of various substances in a frog's blood was compared with their concentration in the fluid along the length of a tubule. This revealed how the swollen inner end of the tubule filters small molecules such as urea, ions, and water from the blood, and how the composition of the fluid is modified as it flows down the tubule, before being expelled into the bladder as urine.

Since these and other early experiments, physiologists have been able to show that the same basic processes discovered in the amphibians also occur in reptiles, birds, and mammals, including ourselves. Modern understanding of human physiology owes a great deal to frogs, and of course the implications of this for medicine can hardly be exaggerated.

A discovery in the early 20th century with profound social consequence came from an experiment on the African clawed toad, *Xenopus laevis*. An extract from the pituitary gland of an ox had been injected into a toad, and within a few hours the toad began to lay eggs. It was soon realized that the toad's ovulation was due to a reproductive hormone called chorionic gonadotropin produced by the ox, and furthermore that the same hormone is present in the urine of pregnant women. From the middle of the last century, injecting a sample of a woman's urine into a female *Xenopus* was widely used as the only reliable pregnancy test. As one doctor wrote: 'Thank you for your report on the pregnancy test on Mrs. X. You may be interested to know that of one GP of many years' standing, one specialist gynaecologist and one frog, only the frog was correct.'

Of all these examples of the role frogs have played in science, none is surely greater than the use of frogs' eggs and embryos in unravelling the mysteries of embryonic development from a single fertilized egg cell to a complete animal form. Amphibian eggs are large and not enclosed in a shell or tough membrane. The black pigmented developing embryo inside is easily observed through the jelly coating. They are also easily harvested in large numbers by injecting the reproductive hormone, as we have just seen. Once collected and fertilized by male sperm, the eggs can be manipulated under a low-power binocular microscope, using instruments little more sophisticated than fine needles and forceps. For example, if just one cell from the early two-, four-, or eight-celled stage is extracted undamaged from the rest and allowed to develop, it still becomes a complete embryo. This led to important conclusions about how the nucleus of the cell controls development. Other work revealed a fundamental process called induction, in which one part of a developing embryo uses molecular messages to bring about appropriate changes in other parts. Most of this early development research was conducted using the African clawed toad, *Xenopus*. Today we know the complete sequence of all the DNA molecules that make up its genetic material. Armed with this information, *Xenopus* continues to be one of a small number of intensively studied species of animals used by molecular specialists interested in how the genes and their environment control the amazing, complex set of processes that turns a single fertilized egg cell into a full-blown animal.

Another aspect of developmental biology that owes much to amphibian research is called regeneration. If a limb of an axolotl, the giant larval salamander, or of an adult newt is amputated, the animal suffers little harm and the lost limb gradually regrows over the next few weeks. Cells move from the body to the stump region, and there they differentiate into the various kinds of specialized cells, such as skin, skeletal, and muscle. From these experiments, much has been discovered about how regeneration takes place. For example, the large white cells of the animal's immune system were found to be essential for successful regrowth. They have also helped us find out about the underlying molecules that control the growth and the differentiation into various types of cells as the new limb develops. The hope is that through the use of amphibians on which experiments can easily be performed, we will discover how to regenerate amputated human limbs, and even how to grow new nerves in place of those damaged in cases of paralysis.

Chapter 8
Conservation and the future of amphibians

Extinction of species of animals and plants has been happening throughout the history of life on Earth. It is the other side of the coin to the origin of new species, for without the loss of the old there would be no opportunity for the new to become established. It is a part of the overall evolutionary process by which all the organisms living today came to be. Furthermore, the fossil record shows that occasional mass extinctions occurred, in which the majority of the existing species disappeared over the course of what, to geologists, is a very brief period of time. There have been five large mass extinctions, the most devastating being that of the end Permian, 250 million years ago which, as we discovered earlier, greatly altered the course of amphibian evolution. After each crisis, the world's biota eventually recovered, thanks to the evolution of many new species from those that were fortunate enough to survive.

Today we find ourselves living through what is often called 'the sixth mass extinction', with many species of plants and animals disappearing around us. Perhaps we should simply accept that this is a natural part of the evolution of life, which we can do little about. But no, the present mass extinction differs from the earlier extinctions in a number of ways. First, although ecologists calculate that relatively few species are disappearing every decade, this is still a far higher rate than during previous mass extinctions, which typically occurred over the course of tens of thousands of years. Secondly, the present extinction is caused directly or indirectly by human activities. Thirdly, we cannot be the least bit confident of eventual recovery by the evolution of new species, as long as *Homo sapiens* continues to exist, expand, and behave as at present.

Amphibians are small, permeable-skinned animals dependent on a humid atmosphere, which makes them especially vulnerable to changes in habitat and climate. They are also sensitive to increasing pollution levels, although there is a great deal of variation in how badly different species are affected; what is a fatal level for some species may scarcely affect others. Most amphibians face a double threat because they rely on two habitats, a more terrestrial one such as forest for the adult, and a body of freshwater for the larvae. Add to this that amphibians continue to be a source of food, medicine, and pets for the rapidly growing human population, and it is no wonder that as a group they face a particularly severe threat. Many species are endangered or threatened with extinction, and for quite a number it is sadly too late, as they have disappeared already.

Amphibians in the wild today

Naturalists around the world agree that the number of amphibians encountered in nature is a great deal lower than it was even a quarter of a century ago, and many would note that some species once familiar to them have disappeared altogether. We met the gastric brooding frogs in Chapter 3. With their habit of nurturing the young in the stomach, these one-time dwellers of the Queensland rainforest were amongst the most extraordinarily adapted amphibians of all. As far as humans are concerned, their story is brief. Discovered in 1972, the last specimens to be seen in the wild were in the mid-1980s, when they died out because of a frog disease introduced into their habitat. almost certainly by humans. There are other similar cases, like the beautiful little viviparous golden coqui that was discovered in Costa Rica in 1986, but is now probably extinct. Amongst urodeles, the tiny Atoyac minute salamander was last seen in the 1980s in the Sierra Madre del Sur of Mexico, after its habitat was largely converted into coffee plantations.

Let us look at the figures from the International Union for Conservation of Nature (IUCN), the most important body concerned with protecting the world's natural inheritance. Its famous Red Book aims to list all the species of animals and plants in the world, with an assessment of how endangered each one is. Every individual species is put into one of a set of categories, from 'Extinct' or 'Extinct in the Wild', to 'Critically Endangered', 'Endangered', 'Vulnerable', and 'Threatened'. The final category is 'Least Concern', for those species not at the moment in any apparent danger of extinction. There is also a category 'Data Deficient' for the considerable number of species where not enough is known about them to be able to make an assessment. An important role of the IUCN is to encourage and coordinate research by field biologists to acquire up-to-date information. To this end, specialist groups working on particular taxa have been set up, one of which is the Amphibian Specialist Group. Thanks to its work, we have a good idea of how amphibians are really faring, and it does not make for cheerful reading. The Red Book tally for 2020–1 is that 34 amphibian species have completely disappeared and 2 no longer occur in the wild. Then there are no fewer than 587 species deemed to be Critically Endangered, and a considerable number of these are almost certainly extinct already. Adding in the species in the Endangered, Vulnerable, and Near-Threatened categories, and assuming that a significant number of the 1,384 Data Deficient species should also be in one of the risk categories, we find that 40–50 per cent of the world's total amphibian species are 'a matter of major concern' to conservationists. Every continent and every kind of habitat is affected.

Natural selection ensures that every species that persists is well adapted to survive in the particular conditions of its habitat: the range of temperatures and humidity, the food available, the predators to be avoided, the season when reproduction is most successful, and so on. We can always say, in broad terms, that a decline in the population, and eventually the extinction of a species, must be due to a change in its habitat in such a way that it is no longer well enough adapted. But there are endless particular ways in which an environment can change to the detriment of a species, and it is usually difficult to pinpoint a single cause. A change in average temperature, for example, might directly affect the development of the tadpole, and also the adult's ability to keep its water balance at the right level. And it is likely to alter other things, such as the plants that give shelter, the food that is available, the competitors for resources, and the level of susceptibility to disease. But despite the complication of multiple factors involved, we can recognize the main causes that are playing a part in the decline of the amphibians.

Loss of habitat

Physical destruction and fragmentation of the habitat by human activities is the single most important cause of the decline of the world's fauna and flora in general, and amphibians are certainly no exception. Every hectare of forest that is felled to plant palm oil trees or to graze cattle, and every pond, stream, and marsh drained to build houses, is more loss of living space for the amphibian inhabitants. Madagascar is a typical example. When humans first arrived on the island 10,000 years ago, a large proportion of the land was covered in tropical forest. But this has been gradually clear-felled over the centuries for small farmers to grow their crops and graze their livestock, in order to feed an ever-expanding population. Now only 10 per cent of the original forest remains, and while most of this remnant is protected within national parks, some is still being lost annually. As for the effect on the amphibian fauna, the most recent conservation assessment shows how serious is the decline in their numbers. Of the 365 species recorded, scarcely more than one-third, 134, were given

Least Concern status. At the other end of the scale almost a quarter, 86, were assessed as Critically Endangered or Endangered. Not enough information was available to assess 14 species, and the remaining 131 were split between Near Threatened and Vulnerable.

England is a very good example of the effect of habitat loss in a temperate region of the world. Never very rich in amphibian species anyway because of its high latitude and isolation from continental Europe, a large decline in the size of their populations began during the Second World War, when large tracts of semi-wild land were dug up to grow food. This continued after the war and up to the present day, with development of intensive cereal farming, and the loss of forests, meadows, and wetlands for new towns and suburbs. Over the past thirty years, three-quarters of the natterjack toads, and around two-thirds of the common toads and great crested newts have been lost.

Amphibian disease

Beginning in the 1970s, naturalists and ecologists in several parts of the world started to notice a rapid decline in the numbers of amphibians that became progressively worse through the next decade or so. It was put down to the usual causes of habitat destruction and pollution from pesticides that were known to be affecting other kinds of organisms, such as mammals and birds. Furthermore, it was assumed that amphibians were more vulnerable to these hazards because of their permeable skin and dependence on humid conditions. However, there were certain features in the decline of the amphibians that were not so easy to explain. One was that the decline was much more severe than that of other groups of vertebrates. Another was how geographically variable it was, with the worst affected areas being the tropical regions of South and Central America and Australia. But the most inexplicable thing was that the decline seemed to be occurring just as severely in well-protected places, such as national parks and nature reserves, where human influences on the habitat and environment were small. For example, by 1981 eight of the thirteen species of frogs in the Reserva Atlantica of Brazil had disappeared. Similarly in Australia, most of the species of amphibians in the very well-protected Queensland national parks were declining and several completely disappeared during the 1980s.

The cause was finally identified in 1998 as a pathogenic fungus with the tongue-twisting name *Batrachochytrium dendrobatidis*. Usually called simply *Bd*, it causes chytridiomycosis, a skin disease of amphibians. Spores of the fungus are carried in water and penetrate the naked skin. Here they develop a network of fine filaments that feed by absorbing the skin's keratin. The fungus grows in size, and produces new infective spores that are discharged back into the water to infect more individuals. The symptoms of a *Bd* infection include thickening of the skin preventing it from functioning properly, loss of pieces of dead skin, and lethargy. Death usually follows within days. While this is far from the first time a fatal infectious disease has caused serious reduction of an animal population, chytridiomycosis is the worst case of pathogenic extinction we know of. It is highly contagious and, unlike most diseases, it affects a very large number of species. So far it seems to have lost little if any of its pathenogenicity, and it persists in the environment, thanks to a number of amphibian and certain other species that act as asymptomatic carriers.

A recent joint study by worldwide amphibian experts calculated that at least 501 species have suffered significant decline due in large part, if not completely, to the fungus. Of these, 90 are probably extinct. The worst affected are amphibians living in wetter conditions because of the water-borne nature of the fungal spores. Amphibians of larger body size also tend to suffer a more severe decline. On the other hand, some species are less severely affected or even immune to the disease. The Californian mountain yellow-legged toad, for example, produces an antimicrobial substance in its skin which prevents the development of the fungal spores. Others seem to develop an immunological resistance after exposure, and this may explain why the populations of a number of badly affected species are now recovering, although none has yet returned to anywhere near the original level.

Asian amphibians are virtually unaffected by the disease, which leads scientists to believe that *Bd* originated there and that the indigenous amphibians have had time to evolve immunity. Furthermore, DNA study of museum specimens proves that the fungus was present in North America by the late 19th century, but it has relatively little effect in that part of the world today. Why then did the disease spread from the northern continents to South and Central America and Australia during the last two or three decades of the 20th century, and have such a devastating effect? The most likely explanation is that infected amphibians were transferred by humans to these regions, and the indigenous species have not had the time to evolve resistance. The spread may have been indirect, as long distance air travel increased. Or it may have been directly due to infected individuals introduced through expansion of the international trade in amphibians for food and pets. It is striking that two species that happen to be immune to chytridiomycosis but are able to carry it are the American bullfrog and the African clawed toad. Both became globally distributed, the former for food, the latter for pregnancy testing and research. The cane toad introduced into Australia is also an immune carrier of the disease, and acts as a reservoir of the fungus in that part of the world.

A number of other amphibian diseases contributing to the worldwide decline have since been identified. In 2016, it was noticed that almost all the fire salamanders in a part of Holland had suddenly died. The culprit was soon discovered to be a similar fungus, *Batrachochytrium salamandrivorans*, to the one killing frogs. *Bsal* as it is known as also came from Asia, and is almost 100 per cent fatal to European salamanders, creating an extremely serious situation. Thirty years ago, an altogether different kind of pathogen caused a sudden mass mortality of frogs around ponds in southern England. It was discovered to be one of the family of ranaviruses, which occur worldwide and can affect fishes and reptiles too. The amphibian ranavirus affects toads and newts as well as frogs, but populations often recover, as some individuals develop immunity to the disease.

Climate change and pollution

The relationship between amphibians and water dominates their life. A change in the presence of water or atmospheric humidity in the environment is likely to have an effect on them, and a warming climate is therefore bound to have consequences. The south-eastern Australian corroboree frog is an example of the effect on the life cycle of abnormal climatic conditions. The male digs a nest in the ground amongst the grass and moss, and the female lays her eggs in it for him to fertilize. The tadpoles soon hatch, but cease to develop any further until the autumn rains come. Predictably enough, the twelve-year drought in Australia starting in 1997 coincided with a severe reduction in the corroboree frog population. But when exceptionally early rains arrived in 2011 and 2012, this also had a severe effect, because now the tadpoles were not ready for it and drowned. Another example is the highland forests of Costa Rica. This is a region rich in amphibians, because they benefit from the mist that envelops the mountains for much of the time. An effect of the warming climate is a reduction in the mist, and falls in the numbers of the land-breeding rain frogs and other amphibians that live in the mountains of Costa Rica, coincided exactly with unusually dry years in which there were the fewest misty days.

The permeable skin of amphibians is very sensitive to pollution in the water and the atmosphere, and the effects of particular pollutants at realistic concentrations is easily tested in the laboratory, unlike the more nebulous factors associated with climate change. Agricultural pesticides generally have damaging consequences at some level. Atrazine is a widely used herbicide that affects the hormones of amphibians, causing males and tadpoles to develop more female traits. It is still a matter of controversy whether the currently allowable level of atrozine in the USA and Australia is harmful to amphibians, although it is now banned in Europe. Nitrogen fertilizers at the levels frequently found in water affect the growth of tadpoles. Heavy metals, such as aluminium, lead, and mercury at levels caused by mining operations can easily reach levels fatal to amphibians, and the effect is even greater when accompanied by acidic conditions.

Invasive species

Movement of people across the globe inevitably causes dispersal of many animals and plants, and introduced species can so disrupt the existing ecosystem that the native species are badly affected. Usually the introduction was accidental, such as bull frogs escaping from farms and not only competing directly with the indigenous amphibians, but also carrying the chytridiomycosis fungus to parts of the world such as South America where it had been absent.

In other cases, the introduction was quite deliberate if well intentioned, such as stocking waters with salmon and trout for sport. We have already mentioned the dire effects on native amphibians of the introduction of the cane toad to eat beetle pests in the Australian sugar cane field. The mosquitofish is a small, 2–3 cm long fish native to the Mississippi Basin and other parts of eastern North America. It has been introduced into several parts of the world to help control mosquito larvae, but of course it eats many other small creatures as well, including tadpoles. In southern California, it has been responsible for a decline in the California newt and the Pacific tree frog. In Australia it has become a major pest, spreading to almost all the states and all kinds of freshwater habitats, with a highly damaging effect on anuran tadpoles.

What must be done: conservation of amphibians

What can be done about the half of the world's 8,000 plus amphibians seriously threatened with extinction within the next few decades? Identifying the largely human-caused hazards to amphibians today—habitat loss, collection for medicine and pets, spread of diseases, pollution, and climate change—is the first step in deciding how best we might be able to save them. Most cases of severe decline of any particular amphibian species are due to a combination of causes. The severely depleted poison dart dendrobatid frogs of South America have suffered from loss of their rainforest habitat as it gives way to palm oil plantations and cattle ranching, from over-collection for the pet trade of such brilliantly coloured animals, from a warming climate disrupting the breeding cycle, and from chytridiomycosis infection. In Europe and North America especially, massive use of agricultural herbicides and pesticides is always a further complicating factor.

The majority of the world's animals and plants are suffering a reduction in population for the same reasons. But amphibians are particularly vulnerable because of their nature and habits; the number of threatened species is around the same as those of the birds and mammals put together. Like the canary in a coal mine, amphibians are often the first to warn that all is not right, and by the end of the last century, it was clear to everybody interested in amphibians that indeed, all was far from right. A worldwide conference, called the Amphibian Conservation Summit, met in 2005 and two years later the IUCN published its *Amphibian Conservation Action Plan*. Updated online in 2015, this is a broad-fronted approach to identifying and proposing solutions to the problem, and work is steadily under way, although there are formidable if not insurmountable problems to face.

The fundamental solution is clearly rigorous protection of large enough areas where the destruction of habitats, and the exploitation of animals and plants, are prevented. Here the amphibians among all the other organisms can play their role as part of the natural ecosystem. These are the National and Transfrontier Parks, and Nature Conservation Areas, which have actually been increasing in total area. They now measure around 12 per cent of the world's surface, although by no means are they all equally well policed, and many necessarily have a human population inside the borders. The second clear need is to halt the rate of climate warming, a problem that is finally starting to be addressed at government level, although much more a response to welcome commercial and public pressure as to the particular needs of amphibians of course.

At least in principle, a more simple need is to drastically reduce the exploitation of populations of amphibians in the wild. With the encouragement of imposing severe penalties for taking wild specimens, the needs of the medical, food, and pet markets can be met by amphibian farming.

But time is not on the side of the many amphibians on the brink of extinction. For them, the best hope of saving individual species is collecting live specimens, breeding them in artificial surroundings, and hoping to be able to reintroduce them in the future into suitably managed areas in the wild. With exceptions, amphibians are actually amongst the easiest vertebrates to breed in captivity because of their small size, broad diet of invertebrates, and external fertilization. Zoos and aquaria are well-placed institutions to take on breeding programmes, and there are plenty of success stories already. The iconic giant salamanders are suffering badly from draining or silting up of their streams. Asa Zoo in Japan is actively breeding giant Japanese salamanders, and Detroit Zoo has a programme to breed hellbenders and eventually reintroduce them into restored streams. The Mallorcan midwife toad is an interesting case. We only knew about it from fossils, and it was believed to have gone extinct centuries ago due to the introduction by humans of predatory snakes and a competing frog. But amazingly, a few specimens were discovered in a remote mountainous region of the island. Some of these were collected and successfully bred. In 1989, seventy-six tadpoles were released back into carefully selected locations, and more in the following years. The toad bred successfully in the wild to the extent that the population soon grew to around 2,000 individuals. Similarly, the North American leopard frog was once believed to be extinct, until some were rediscovered in the remote desert springs of north Nevada. These were bred and have been successfully reintroduced into the wild. One of the most spectacular species to have been bred in captivity is the brilliant blue poison dart frog of Suriname, highly popular as a pet.

In most cases reintroduction poses its own problems. Often there is simply no suitable, large enough habitat available for a successful release. There is also the likelihood of reinfection by disease where that had been a major part of the original decline. Success in the future may depend on a more sophisticated approach, such as genetic manipulation of individuals to make them more disease resistant. One novel approach tried recently concerns the Montserrat mountain chicken frog. A population was released into a semi-wild enclosure in which the small ponds were warmed by solar power to a high enough temperature to inhibit the fungal pathogen. In other cases, attempts are being made to find and introduce pathogens that kill the fungus itself. But for all its potential, it is unrealistic to expect that all 3,000–4,000 species of amphibians in urgent need of attention can be saved this way.

Cloning is an even more radical possibility for saving an individual species. Gastric brooding frogs became extinct during the 1980s, but fortunately some tissue had been preserved in deep freeze. Researchers in Australia replaced the fertilized nuclei of the eggs of a related species of frog with thawed-out gastric brooding frog nuclei. The eggs were incubated in the laboratory, and some of them developed up to a multicellular stage. Although far from successful so far, this is enough to show that in principle recovering a completely extinct species in this way is possible.

So what is the realistic outlook for the world's amphibians, these amazing, beautiful, unique creatures? Not so long ago the answer would have been: pretty poor. The human population is expanding at a prodigious rate, and governments have been obsessed with economic growth at the expense of all else. But straws in the wind are accumulating. There is no longer any serious scientific doubt that human-induced climate change and pollution are ultimately as much to the detriment of ourselves as to amphibians and the rest of the natural world. As for the loss of biodiversity, with the help of respected educators, entertainers, and communicators, it seems that more and more people do not wish it to happen, and are increasingly prepared to pressurize their governments to do what needs to be done.

Further reading

The Amphibia Web. <https://amphibiaweb.org/index.html>

The International Union for the Conservation of Nature Amphibian Specialist Group. <https://www.iucn.org/commissions/ssc-groups/amphibians-reptiles/amphibian>

Halliday, T. and Adler, K. 2002. *The new encyclopedia of reptiles and amphibians*. Oxford: Oxford University Press.

Wells, K. 2007. *The ecology and behaviour of amphibians*. Chicago: University of Chicago Press. A large, very detailed and scholarly account.

Pough, F. H. et al. 2016. *Herpetology*, 4th edition. Oxford: Sinauer Associates. Excellent and well-illustrated textbook account.

Halliday, T. 2016. *The book of frogs*. London: Ivy Press. Beautifully illustrated descriptions of about 500 anuran species.

Attenborough, D. 2008. *Life in cold blood*. London: BBC Books. An easy account by the master of natural history communication.

Duellman, W. E. and Trueb, L. 1986. *Biology of amphibians*. New York: McGraw-Hill. A classic account covering all aspects of amphibian morphology, biology, and behaviour.

Clack, J. A. 2012. *Gaining ground: the origin and evolution of tetrapods*, 2nd edition. A detailed, definitive account of the history of the tetrapods.

Carroll, R. L. 2009. *The rise of amphibians: 365 million years of evolution*. Baltimore: The John Hopkins University Press. A very detailed, specialist fossil history.

Index

A

Acanthostega 27, 29, 32
advertisement calls 97
aestivation 24
African clawed frog (*Xenopus*) 2, 4, 6, 21, 77–8, 86, 109, 115–16
aggressive behaviour 99, 100
aggressive calls 97
aïstopods 38–9
alpine newt 91
American bullfrog 50, 77, 108, 125
amniotes 35–6
amphibamids 44
Amphibamus 43, 44
Amphibian Conservation Summit 126
amphibians
  definition 1
  evolution 41–7
  as food 106–8
  habitats *see* habitats
  medicinal properties 108–10
  as pets 110–11
Amphibian Specialist Group 119
amphiumas (congo eels) 9, 70–1
amplexus 50–2, 60
Andreas scheucheri fossil 106, 107
anthracosaurs 33–6
antibiotics 109
anurans 3–6
  direct development 67
  evolution 41–3
  feeding 86–8
  larvae 54–7
  legs 75–80
  mating 49–52
  metamorphosis 59
  senses 90, 91–8
  viviparous 65–6
Apoda 9; *see also* caecilians
aposematic coloration 13–15
aquatic habitats 6, 8
Archegosaurus 36–8
ascaphids 5
Atoyac minute salamander 119
Australia, endangered species 122, 124, 125–6
axolotl 67–9

B

Balenerpeton 36
balloon frog 79
barred leaf frog 78
batrachotoxins 111–12
Bd (*Batrachochytrium dendrobatidis*) 122–3
Beiyanerpeton 44
bird-poop frog 13, 14

blood vessels 11, 12, 18–19, 20
body language 99, 100
body temperature 22–4
Bombay night frog 51
branchial respiration 16, 18–19
Brazil
endangered species 121–2
medicinal properties of amphibians 109
breeding programmes 127–8
bright-eyed frog 5
bromeliads 63
Bsal (*Batrachochytrium salamandrivorans*) 123–4
buccopharyngeal respiration 16–17
Bufonidae family 3; *see also* toads
bullfrog, American 50, 77, 108, 125
Bunyan, John 105
burrowing
caecilians 81
to escape predators 79
to hibernate 23–4
to prevent water loss 20–1
to protect eggs 61–3
burrowing toad 6
feeding 86–8
burrows 5–6, 10
butterfly display 99

C

caecilians 9–10, 21
diet 89
direct development 67
evolution 41, 43, 44–5
feeding 88–9
larvae 56, 57–8
locomotion 81
mating 54
senses 90, 101–2
viviparous 66
camouflage 13–14
cane toad 57, 86, 112–13, 123, 125
cannibalism 57
canyon tree frog 22–3
Carboniferous Period 31–40
Caudata 6; *see also* urodeles
cave-dwelling urodeles 98
cave salamander 85
China
frogs as food 108
myths and folklore 104
Chinelestegophis 45
Chinese giant salamander 7, 8
Chinese medicine 108–9
chromatophores 11, 13
Clack, Jenny 29
clawed frog (*Xenopus*) 2, 4, 6, 21, 77–8, 86, 109, 115–16
climate change 124, 127
cloning 128–9
coloration 13–15, 93
colour vision, anurans 92–3
common frog 93, 113
communication
body language 99, 100
coloration 93
leg-waving 93, 94
vocalization 96–8
congo eels (amphiumas) 9, 70–1
conservation 126–9
coqui frog correlated progression 30
corroboree frog 124
Costa Rica 124
crab-eating frog 2, 21
Crassigyrinus 35
crawling 78
crests 99
Cretaceous Period 46
cryptobranchids 8, 71, 83, 106
cutaneous respiration 12, 16

D

dancing frog 94
Darwin's frog 62, 64–5
dendrobatids 15
dermis 11, 12
Devonian Period 26–31

diet 82
caecilians 89
tadpoles 57
urodele larvae 57
see also feeding
digestion 55
Diplocaulus 38–40
direct development 66–7
Discosauriscus 36
diseases 121–4, 128
distress calls 97–8
Dumeril, Auguste 67
Dvinosaurus 38

E

ears, anurans 93–5
eastern gray tree frog 13
ecosystem 112–13
ectothermy 22
edible amphibians 106–8
education 113–16
eggs 48–9
laid on land 66–7
protection of 60–5
in scientific research and education 115–16
Egypt, myths and folklore 104
embryo research 115–16
endangered species 118–29
endothermy 22
Eocaecilia 43, 44
epidermis 11–12
Eryops 36, 37, 38
Europe
frogs as food 108
habitat loss 121
myths and folklore 105
European common frog 93, 113
Eusthenopteron 26–8
evolution
modern amphibians 41–7
tetrapods 25–41
extinctions 31–3, 40, 46, 117–19
eyes 91–2, 98

F

facultative paedomorphs 69
fairy tales 105
feeding 82
anurans 86–8
caecilians 88–9
tadpoles 55–6, 88
urodeles 82–5
see also diet
fire-bellied newt 110
fire-bellied toad 15, 110
fire salamander 7, 15, 66, 83–5, 123
fish
evolution 26–8
locomotion 72
threat to amphibians 125–6
folklore 104–6
food, amphibians as 106–8
fossils 26–30, 33, 42
Fowler's toad 78
freezing 23–4
frogs 3
dissection 113–14
evolution 42
eyes 91–2
hibernation 23
legs 75–80
mating calls 49–50, 96
meat 106–8
in myths and folklore 104
in scientific research and education 113–16
skin 12–15
fungus infections 122–4, 128

G

gastric brooding frog 65, 118, 128–9
Gephyrostegus 35
giant salamander 2, 7, 8, 127

gill chamber 54
gills 16, 18–19, 54, 88
glass frog 80
gliding frog 78, 80
golden coqui frog 65–6, 118–19
golden dart frog 111–12
golden frog 5
golden mantella frog 110
golden poison frog 15
goliath frog 4, 108
gray tree frog 13
great-crested newt 99
grey foam-nest tree frog 61
gulping movements 18

H

habitats 2–3, 10–11
  aquatic 6, 8
  burrows 5–6, 10
  caves 98
  damp 19–20
  loss of 120–1, 126–8
  trees 5, 79–80
hairy frog 16
hammerhead salamander 38–40
hearing
  anurans 94–8
  urodeles 98
hellbender salamander 8, 71, 83
hibernation 23–4
Homo diluvii testis fossil 106, 107
Hong Kong 110
hopping 78–9; *see also* jumping
hydration 19–20, 49
Hylidae family 5
hyoid 83, 85

I

Ichthyophis 56, 57–8
Ichthyostega 27, 28–9, 32
Indonesia, frogs as food 108
internal concertina locomotion 81
International Union for Conservation of Nature (IUCN) 119, 126
invasive species 125–6
iridophores 13

J

Japanese giant salamander 7, 8, 127
jaws 88–9
jewel-eyed tree frog 63
jewels 105
jumping 75–7; *see also* hopping
Jurassic Period 46

K

keratin 11–12
kidneys 21

L

Lake Junin frog 6
lamellae 18–19
larvae 48
  anurans 54–7
  caecilians 57–8
  feeding 83
  metamorphosis 49, 57, 58–9
  senses 90–1, 102–3
  urodeles 57, 83
lateral lines 36, 90–1, 102–3
lateral undulation 73–4, 81
leaf litter frogs 4
leaping 75–7; *see also* hopping
legs 42, 73–4
  frogs 75–80
leg-waving 93, 94
lek-breeders 99
leopard frog 113, 128
lepospondyls 38–9, 40, 44
limbs, regeneration 116
lipid glands 13
little bromeliad tree frog 63

lobed-fin fish 26–7, 28
locomotion 72–4
  caecilians 81
  frogs 75–80
lungless salamanders 8, 17, 18, 98
lungs 16, 17–18
  evolution 28
Lydekkerina 40

M

Mackie, William 26
Madagascar 5, 42, 110, 120–1
mantellids 5, 110
marbled newt 91
marble snouted browser frog (shovel-nosed) 61–3, 79
marsh frog 97
marsupial frogs 64
mass extinctions 31–3, 40, 46, 117–18
Mastodonsaurus 37, 40–1
mating 49, 50–4
  displays 99
  frog chorus 49–50, 96
  pheromones 101
  use of colour in 93
  see also reproduction
medicinal properties of amphibians 108–10
Megophryidae family 4
metamorphosis 49, 57, 58–9
Mexican burrowing toad 6
  feeding 86–8
Mexican pygmy salamander 7
Microbrachis 38
microsaurs 38–9, 45
midwife toad 62, 63–4, 127–8
moor frog 93
mosquitofish 125–6
mossy tree frog 13, 14
mountain chicken frog 128
moustache toads 4
mouths 16–17
  tadpoles 55
  see also feeding
mucous glands 11, 12
mudpuppies 9, 19, 69–70
Myobatrachus (turtle frog) 2, 52
myths 104–6

N

natural selection 119–20
navigation 91
nectridians 38–9
nerves 11
nests 61
newts 6, 8
  locomotion 73
  navigation 91

O

obligatory paedomorphs 69–70
occipital condyles 75
olfactory senses 99–101
olm 9, 19, 69–70, 98
operculum 54
orange-sided leaf frog 13
osmosis 21

P

Pacific giant salamander 98
paedomorphosis 8–9, 19, 44, 59, 67–71
painkillers 109–10
palatal fenestrae 36
Panderichthys 27, 28
Pantylus 38–9
paradoxical (shrinking) frog 57, 110
parental care 48–9
  direct development 66–7
  protection of eggs 60–5
  viviparity 65–6
Pederpes 33, 34
Pelobatidae 5–6

Peltobatrachus 38
pelvic patch 20
Permian Period 36, 40
pesticides 125
pets, amphibians as 110–11
pheromones 52, 54, 99–101
phyllomedusan tree frogs 56
Pipa (Suriname toad) 4, 6, 18, 64, 86
pipids 4, 6
pithing 114
plants, evolution 26, 46
plethodontids 8, 17, 18, 52–3
  direct development 67
  eyes 98
  feeding 84–5
poison dart frog 15, 109, 111, 126, 128
poison darts 111–12
poison glands 11, 12, 15
pollution 124–5
pouched frog 64
pregnancy tests 115
Prosalirus 42
prostaglandin 65
Proteidae family 8–9
Proterogyrinus 33–5
pulmonary respiration 16, 17–18
pygmy salamander 7

Q

Quito stubfoot toad 56–7

R

rain frog 50–1, 65–6, 67
ranavirus 124
Ranidae family 3; *see also* frogs
red-backed salamander 99, 100
Red Book IUCN 119
red eft 15
red-eyed tree frog 51, 60, 80, 111
red-legged salamander 52–3
red-spotted newt 91
regeneration 116
reintroduction 128
reproduction 49, 50–4; *see also* mating
reptiles 35
research 113–16
respiration 15–16
  branchial 16, 18–19
  buccopharyngeal 16–17
  cutaneous 12, 16
  pulmonary 16, 17–18
  tadpoles 54
retractor muscle 85
Rhacophoridae family 5
rhipidistians 26–8, 30, 72
Rhynchonkus 45
Rhynie Chert, Scotland 26
robber frog *see* rain frog
Romer, Al ('Romer's Gap') 33
running 73

S

salamanders 6, 7–8
  direct development 67
  evolution 38–40, 44
  feeding 82–5
  in folklore 106
  habitats 2
  hibernation 23–4
  larvae 55
  locomotion 73–4
  mating 52–4
  paedomorphosis 67–71
  senses 98–101
  skin 15
  viviparous 66
Salientia 3; *see also* anurans
salt levels 21, 114
Säve-Söderburgh, Gunnar 28
Scaphiopodidae 5–6
Scheuchzer, Johann 106
scientific research and education 113–16
seawater 2–3, 21, 25

senses
  anurans 90, 91–8
  caecilians 90, 101–2
  lateral lines 36, 90–1, 102–3
  urodeles 90, 98–101
Seymouria 36
Shakespeare, William 105
sharp-nosed frog 77
shovel-nosed frog (marble snouted browser) 61–3, 79
shrinking (paradoxical) frog 57, 110
Siberian salamander 24
sirens 2, 8, 19, 21, 69–70
skin 11–12
  coloration 13–15
  permeable 2, 8, 12, 16, 20
  poisonous 111–12
  surface area 16
  water levels 19–21
smell, sense of 99–101
spadefoot toad 5–6, 20, 49, 58, 78
Spain, medicinal properties of amphibians 108
spermatophores 52–3
spinal column 75
spiracle 54–5
spotted toad 49
stereospondyls 37, 40–1, 45
strawberry poison frog 63
sunlight, basking in 22
Suriname toad (*Pipa*) 4, 6, 18, 64, 86
swimming 73–4, 77–8, 81

T

tadpoles 48, 49, 54–7
  colour vision 92–3
  feeding 88
  locomotion 74
  metamorphosis 59
tailed frogs 5, 51–2
tails 54
Taita Hills caecilian 67
teeth 88–9
temnospondyls 36–8, 40, 45
temperature regulation 22–4
temple tree frog 60–1
tentacles 101–2
tetrapods 25–41
  locomotion 72–3
Texas blind salamander 98
thyroxine 58
tiger salamander 67–9, 83–5
Tiktaalik 27, 28
toadfrogs (Megophryidae family) 4
toads 3
  in myths and folklore 104–5
toadstones 105
tongues 84–5, 86–7
torpor 22, 24, 49
tortoise frog *see* turtle frog (*Myobatrachus*)
Tournier's forest toad 66
toxins 111–12
tree frogs 13, 46, 56
  grey foam-nest 61
  jewel-eyed 63
  little bromeliad 63
  locomotion 79–80
  phyllomedusan 56
  temple 60–1
  warning calls 98
trees 5
Trematosaurus 37, 41
Triadobatrachus 42, 43
Triassic Period 40–1
túngara frog 62
turtle frog (*Myobatrachus*) 2, 52
Typhlonectes 81

U

urea 20
urodeles 6–9
  diet 83
  evolution 41, 44
  feeding 82–5
  in folklore 106
  larvae 57, 83
  locomotion 73–4

urodeles (*cont.*)
  mating 52–4
  metamorphosis 59
  paedomorphosis 67–71
  senses 90, 98–101
  skin 15
  viviparous 66

V

vertebrae 5, 75–6
Vijayan's frog 4
viviparity 65–6

W

walking 73–4
Wallace's flying frog 78, 80
warning calls 97–8
warning coloration 13–15
water-holding frog 108
water levels 19–21
water temperature 23
web-toed salamander 85
Westlothiana 34, 35–6
Whiteaves, F. E. 26–7
Wood, Stan 35
wood frog 2, 23–4

X

Xenopus (clawed frog) 2, 4, 6, 21, 77–8, 86, 109, 115–16

Y

yellow-bellied toad 109
yellow-legged toad 122–3

Z

zoos 127